SAS Data Management for Public Health

An Introduction

SAS Data Management for Public Health

An Introduction

1st Edition

Brianna Magnusson and Caroline Stampfel

Brigham Young University

cognella®

SAN DIEGO

Bassim Hamadeh, CEO and Publisher
Tony Paese, Project Editor
Abbey Hastings, Associate Production Editor
Emely Villavicencio, Senior Graphic Designer
Alexa Lucido, Licensing Manager
Natalie Piccotti, Director of Marketing
Kassie Graves, Vice President of Editorial
Jamie Giganti, Director of Academic Publishing

Cover image copyright © 2018 iStockphoto LP/gremlin.
Interior image copyright © 2012 Depositphotos/shirophoto.

Printed in the United States of America.

3970 Sorrento Valley Blvd., Ste. 500, San Diego, CA 92121

Brief Contents

Contents

About the Data

The data used in most places in this book comes from the National Health and Nutrition Examination Survey (NHANES). NHANES is a continuous cross-sectional survey administered by the United States National Center for Health Statistics (NCHS). You can find complete information about the survey by visiting www.cdc.gov/nchs/nhanes.

In brief, the study was first administered in the 1960s and continues today. The study consists of an interview that includes information on demographic, socioeconomic, dietary, and health-related questions, and is followed by an examination that consists of medical, dental, laboratory, and physical measurements.

The datasets used in this book can be freely downloaded from the NCHS website. You can also find the codebook or data dictionaries for this data at the same website: https://wwwn.cdc.gov/nchs/nhanes/Default.aspx.

Lesson 1

Getting Started with SAS

- ▶ Describe the basic SAS features,

- ▶ Explain the difference between SAS steps and SAS statements,

- ▶ Explain the structure of a SAS dataset,

- ▶ Create a permanent SAS library using a libname statement,

- ▶ Explain the difference between permanent libraries and the work library,

- ▶ Run a PROC CONTENTS and interpret the output,

- ▶ Demonstrate saving a SAS program as a .SAS file, and

- ▶ Demonstrate opening a previously saved .SAS file.

Navigating through SAS

When you first open SAS, there are three subwindows within the program screen: the log, the editor, and the explorer windows.

The Log

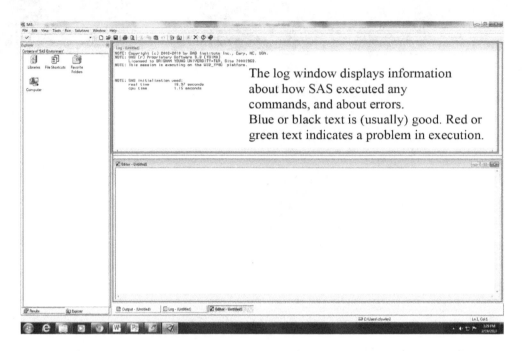

Figure 1.1

The Editor

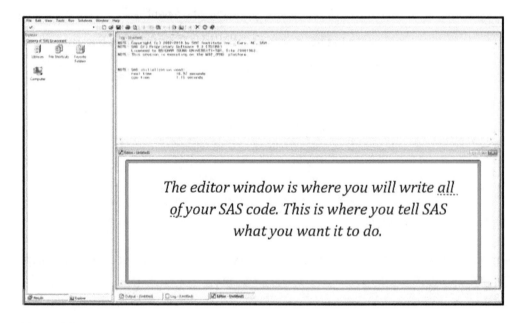

Figure 1.2

SAS Data Management for Public Health

The Explorer

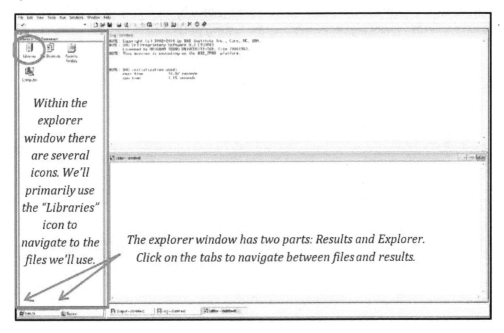

Within the explorer window there are several icons. We'll primarily use the "Libraries" icon to navigate to the files we'll use.

The explorer window has two parts: Results and Explorer. Click on the tabs to navigate between files and results.

Figure 1.3

A fourth window opens in SAS when you have begun to run analyses. This window is the Output or Results window. By default, in SAS versions 9.3 or later, this is a HTML results viewer.

Figure 1.4

In earlier versions of SAS, the default output window is the listing.

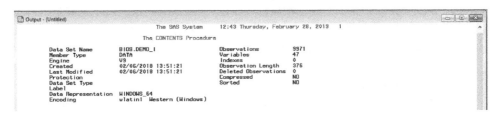

Figure 1.5

Although the two windows look different, they contain the same information. If you prefer one window/output style over the other, you can select your choice by choosing Tools→ Options→Preferences→Results and then checking or unchecking the Listing or HTML boxes.

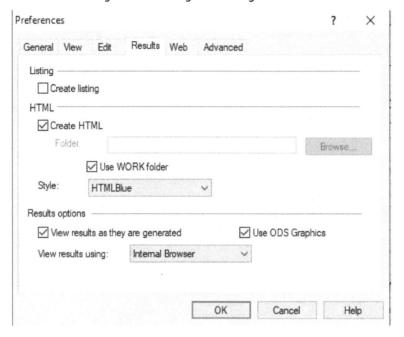

Figure 1.6

The Menu Bar

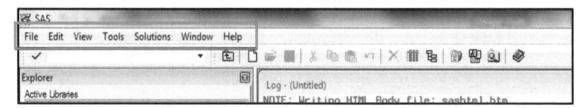

Figure 1.7   As with many software programs, SAS contains a menu bar that can be used to navigate through the options. The menu bar includes several drop-down menus, including File, Edit, View, Tools, and Help. The file menu is important because you can save SAS programs or other files and open previously saved files.

Short Cuts

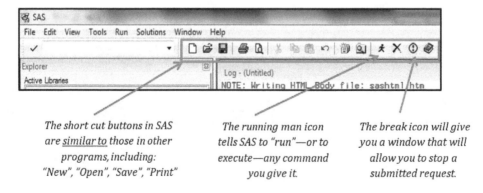

The short cut buttons in SAS are *similar to* those in other programs, including: "New", "Open", "Save", "Print"

The running man icon tells SAS to "run"—or to execute—any command you give it.

The break icon will give you a window that will allow you to stop a submitted request.

Figure 1.8

SAS Statements

All SAS programs are made up of SAS statements. A SAS statement begins with a SAS keyword or a step command and always ends with a semicolon (;). Groups of SAS statements make up the SAS steps.

```
data newdatasetname; set olddatasetname;
```

One SAS statement

Figure 1.9

SAS Steps

SAS has two different kinds of "Steps." Steps are groups of SAS statements that tell SAS to do certain things.

1. DATA steps
2. Procedure (PROC) steps

DATA Steps

DATA steps begin with the step command DATA and end with the step command RUN.

> **Note 1.1 ▼**
>
> The DATA step does not produce any output. A Procedure step (PROC) must be used with or following a DATA step to examine any changes.

Within the DATA step, you can

- create new datasets,
- copy or merge existing datasets,
- clean data to remove missing or invalid values,
- recode variables (e.g., collapse categories), and
- calculate new variables from the values of an old variable.

The basic syntax for a DATA step looks like this:

```
data newdatasetname; set olddatasetname;
/*In this section ("within the data step"), you will type all of
your cleaning, recoding, or data management statements*/
run;
```

Procedure (PROC) Steps

PROC steps begin with the step command PROC and end with the step command RUN; there are many different PROC steps that are used to

- look at the data, and
- run frequencies, means, and a variety of statistics.

Each PROC step does something different. Some examples of PROC steps include the following:

PROC FREQ: calculates frequencies, cross tabulations, chi-square, odds ratios
PROC MEANS: calculates the mean, sum, median, min/max
PROC SORT: sorts data by selected variable(s)
PROC CONTENTS: returns information about your dataset
PROC FORMAT: allows you to create values for your variables
PROC PRINT: allows you to display information from a dataset
PROC LOGISTIC: allows you to build a logistic regression model

We will learn more about each of these PROC steps throughout this book.

SAS Syntax

The code you will write is known as syntax. There are several important things to know about SAS syntax.

The Semicolon

The semicolon is the most important punctuation in SAS. Every SAS statement ends with a semicolon. Forgetting to include a semicolon is the most common type of error in SAS programming. If your log shows errors or your program is not executing the way you think it should, check your semicolons first.

Text Colors in SAS

Within the editor window, SAS uses a variety of colors for different types of commands. These colors can be an important clue to helping you construct and troubleshoot code.

- Dark Blue = Step commands (DATA, PROC, and RUN)
- Light Blue = SAS operators such as SET, IF, THEN, ELSE, and options for DATA steps or PROC steps
- Green = Commented text. Commented text are notes you make in the SAS program that do not run when the program runs (i.e., SAS doesn't "see" them)
- Black = Variable names, dataset names, etc.
- Teal Green = Format names and numbers (Arabic numerals)
- Purple = Titles or other things between quotes in your SAS program
- Red = An error. Always make sure you have ended your SAS statement with a semicolon

Rules for SAS Statements

As we proceed through the lessons, you will note that some things in SAS will be capitalized and others will not. You'll also note that lines of code will be organized in groups and otherwise delineated with spaces, comments, or indentations. SAS does not care about capitalization or the spacing of lines of code. I will usually show each SAS statement on its own line; this is a convention that I, and many other SAS programmers, use to aid in the organization of code and to help with troubleshooting. This is not required by SAS. You can and should find your own conventions. There are only a few rules for SAS statements:

1. SAS statements end with a semicolon.
2. You can enter SAS statements in lowercase, uppercase, or a combination of the two.
3. You can begin SAS statements in any column of a line and write several statements in a line (separated by semicolons, of course).
4. You can begin a statement on one line and continue it on another line, but you cannot split a word between two lines.
5. Words in SAS statements are separated by blanks or by special characters (such as the equal sign and the minus sign in a calculation).

SAS DataSets

At its core, SAS is a data management system. Databases are the heart and soul of SAS, because without them, no analysis or graphing would be possible. We'll begin by looking at datasets that are already "SAS datasets"; those that are already formatted for SAS use (ending in the .sas7bdat file extension). In upcoming lessons, we will look at how to take data from other formats, such as text documents, excel files, and datasets from other software programs and import them into SAS.

Data Structure in SAS

SAS data files are essentially large spreadsheets. Each column represents one variable or piece of information and each row represents a single observation, typically one person's data for each variable. Variable names are identified in the column headers.

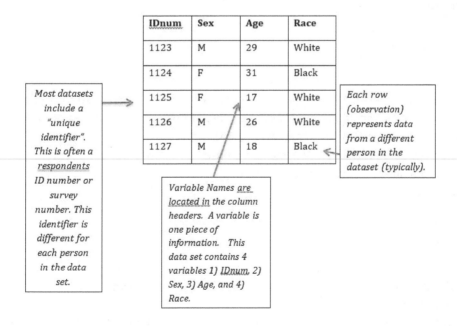

Figure 1.10

Accessing Data in SAS

Most datasets used in public health research are extraordinarily large, containing hundreds or thousands of variables, and hundreds or thousands of observations. It is usually not possible to adequately assess the content of any database that large by simply looking at the spreadsheet. SAS uses a variety of DATA and PROC steps to manipulate, view, and analyze data. First however, we must open our datasets in SAS. We will do this using SAS libraries.

SAS Libraries

A SAS library is a pointer or shortcut to a folder in which you have saved the datasets you wish to work with. This is similar to an icon on your computer's desktop. SAS has two types of libraries: permanent and temporary. Permanent libraries save data files on your computer or flash drive, making these files available to you during and after your SAS session. Temporary libraries save data files to a temporary folder in SAS that empties every time you end your SAS session. Both temporary and permanent libraries have their place. As a rule of thumb, use permanent libraries for datasets you want to keep and temporary libraries for those that you do not want to keep. Keep in mind that SAS datasets can be very large (many kilobytes) so you'll generally only want to keep copies of datasets you need to use again.

Permanent Libraries

Permanent libraries can be created using one of two methods: (1) the libname statement and (2) the point-and-click method. Permanent libraries have names you create. For example, I could create a permanent library called NHANES to store all of the data from an NHANES survey. **Library names can include letters, numbers, and underscores, but no other special characters, and they cannot exceed eight characters**.

The Libname Statement

The libname statement is a SAS statement that creates the pointer (shortcut) to the folder on your hard drive or removable media where your datasets are stored. This does not create a new folder; it just assigns an "abbreviation" to the filepath so that SAS can easily find your datasets. Then this "abbreviation," called the libref or the library name, can be used within the SAS editor to refer to the data.

The syntax for a libname statement looks like this:

```
libname libref 'filepath';
```

- Libname = the SAS keyword (light blue).
- Libref = the name of your library (letters and numbers, no special characters, and less than or equal to eight characters).
- Filepath = where the folder containing the datasets is saved on your computer or flash drive or in the cloud.
  - Example Filepath = L:\440\DataSets

To get to the folder that contains my datasets on my flash drive, I would use the following syntax:

```
libname nhanes 'L:\440_Fall 2012\Data Sets';
```

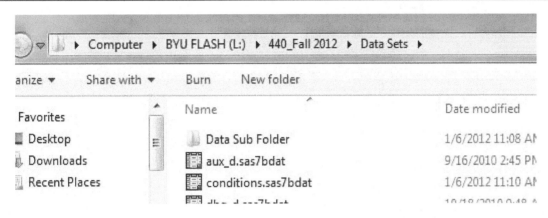

Figure 1.11   You can get the filepath by navigating to the folder where your datasets are stored and then clicking in the address bar and right-clicking to copy the filepath.

Check the log. After creating a library using the libname statement, your log should tell you that the library was successfully assigned.

```
13    libname nhanes 'L:\440_Fall 2012\Data Sets':
NOTE: Libref NHANES was successfully assigned as follows:
      Engine:         V9
      Physical Name: L:\440_Fall 2012\Data Sets
```

Figure 1.12

If you have an error message, check the filepath. An incorrect filepath is the most common reason for an error when using a libname statement. The error in the SAS log is illustrated in the green text: *Library NHANES does not exist*. Note that it still tells me the library was successfully assigned, but an inspection of the libname statement tells me that the filepath I have specified is incorrect. Comparing the two windows, I can see that in the second window, I have incorrectly specified the filepath as D:\440_Fall 2012\DataSets when in fact that folder is found in disk drive L.

```
WARNING: Library NHANES does not exist.
NOTE: Libref NHANES was successfully assigned as follows:
      Engine:         V9
      Physical Name: D:\440_Fall 2012\Data Sets
```

Figure 1.13

Check the explorer. Navigate through the explorer window to check that your library has been created and contains the datasets you want to use.

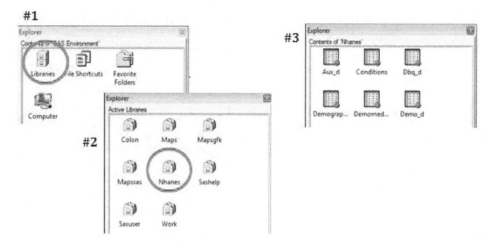

Figure 1.14

Any library you create using a libname statement is permanent, meaning that any dataset you save into the library will be saved on your computer or flash drive (wherever your libref specifies) and will be available during and after your SAS session. *However*, the library itself must be reassigned each time you open SAS. **This means that each time you open a new SAS session, you must run the libname statement to assign the library that contains the datasets you wish to work with.**

Note 1.2 ▼

When storing datasets on removable media, such as a flash drive, the drive letter may change when you switch from computer to computer. For example, in my office, the correct filepath for the NHANES datasets is L:\440_Fall 2012\DataSets; whereas in the computer lab, this changes to D:\440_Fall 2012\DataSets. Please also note that if you change the name of or reorganize pertinent folders on your flash drive, this may also change your filepath.

The earlier example shows how to create a library using a filepath when SAS is installed on your local computer. If you are using SAS Studio, see the appendix at the end of this book for information specific to that application (i.e., SAS University Edition). Some of you may access SAS through a remote server. Locating the filepath from the server is more easily done using the point-and-click method. See the following for an example of how to access SAS using a Citrix remote applications platform.

The Point-and-Click Method—Local Installation of SAS

The point-and-click method uses a menu-driven approach to create a library. As it is menu driven, there is no syntax to write. There is also an option to make this library appear every time you open SAS; unfortunately, this only works if you are always using the same computer. Complete the following steps to add a library using the point-and-click method:

1. Right-click in the explorer window.
2. Select "New."

3. Double-click on the "Library" icon (or push the libraries button on the menu bar).
4. Type in the libref you have chosen.
5. Press the browse button and navigate to the folder where you store your datasets.
6. Check "enable at startup" to have this library open each time you open SAS.
7. Click OK.

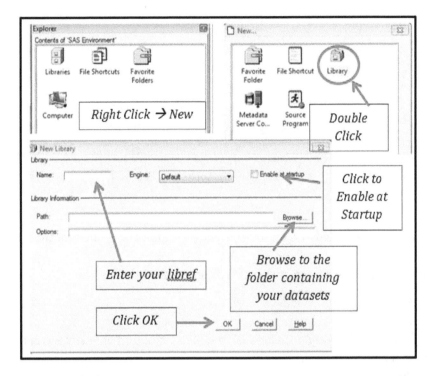

Figure 1.15

   Enable at startup will automatically open this library each time you open SAS on that computer. This is particularly helpful if you are primarily working on the same computer day after day, but it is not useful if you regularly change machines.

The Point-and-Click Method—Citrix Remote Applications Platform

Many universities and workplaces use a server to house a program such as SAS. This is a useful tool, as it allows many people access to the programs without having to install programs locally. Complete the following steps to create a library using the point-and-click method using a remote applications platform.

1. Go the web address for your remote applications platform. Note that you make need to install a receiver or driver onto your local computer to enable the platform.
2. Open SAS.

Once I have gotten this far, SAS looks the same as it does when I'm using the desktop version.

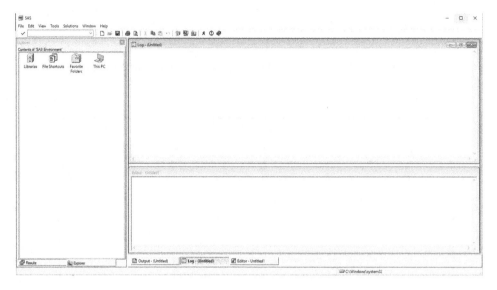

Figure 1.16

Now I can use the same steps I used before with the point-and-click method to set up my permanent library:

To add a library using the point-and-click method complete the following:

1. Right-click in the explorer window.
2. Select "New."
3. Double-click on the "Library" icon (or push the libraries button on the menu bar).
4. Type in the libref you have chosen.
5. Press the browse button and navigate to the folder where you store your datasets.
6. Check "enable at startup" to have this library open each time you open SAS.
7. Click OK.

What changes is how my computer files look when I browse for them.

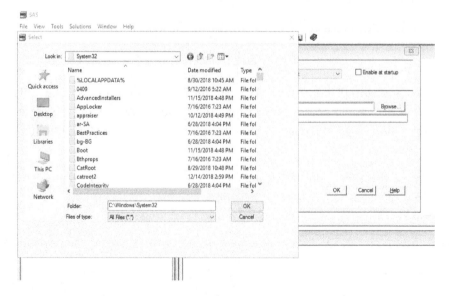

Figure 1.17   Notice that when I select browse, I see a variety of files that look different than what I ordinarily see. I need to click through several of these windows to locate my local computer.

To find my local computer from here, I need to click through several steps. For my system, this is THIS PC → LOCAL DISK (C:)

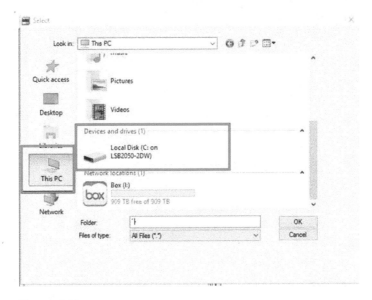

Figure 1.18

Then Users → My username for the computer

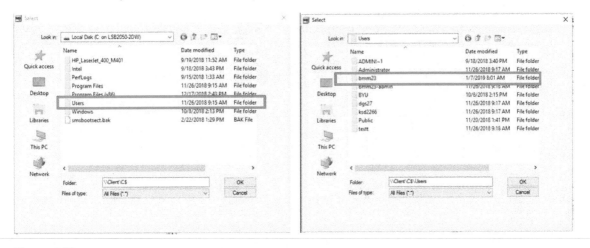

Figure 1.19

The next window should then look like the file structure you are accustomed to seeing on your computer.

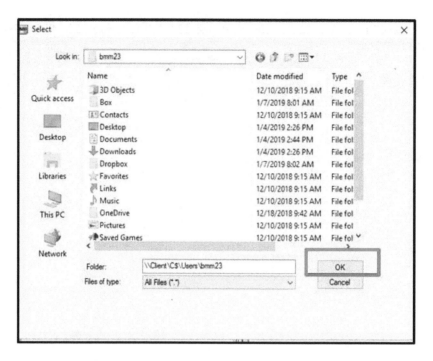

Figure 1.20   From here, I can continue to navigate through my documents, my cloud storage locations (Dropbox or Box, etc.), and the file where I have saved my data files.

Once you have located the folder containing your data files, select OK and then OK in the SAS window.

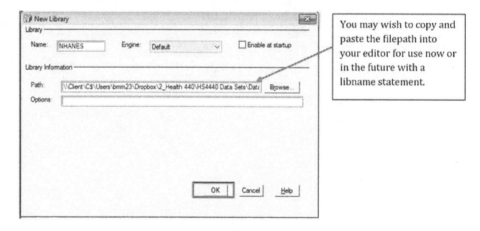

You may wish to copy and paste the filepath into your editor for use now or in the future with a libname statement.

Figure 1.21

Temporary Libraries

The temporary library, also known as the "work" library, is automatically generated by SAS each time the program is opened. Anytime you access data without using a libname statement or the point-and-click method, that data will be stored in the work library. We will later discuss how the work library is used when making copies of data or when using a DATA step to enter data. The work library is very useful, as it can prevent having dozens of copies of the same dataset permanently saved to your computer or flash drive; however, the work library disappears as soon as SAS is closed and will be empty when

SAS is re-opened. It's important to ensure that you have saved a permanent copy of the data or saved your program prior to closing SAS so that you can recreate it if necessary.

DataSet Names

Once you have created your library, you can begin to look at your dataset using PROC steps. In both PROC and DATA steps, you will need to tell SAS which dataset you would like to use. SAS uses a compound (two-part) naming system for datasets.

- Part 1 = libref (library name from the libname statement).
- Part 2 = dataset name (name of the file, without the file extension; e.g., mcq_d.sas7bdat with only the mcq_d).

When you reference a dataset name in SAS, you must specify the two-part name: libref & dataset name, separated by a period: libref.datasetname.

For example, if I had created a library called SURVEY, and my dataset was called UTDRIVERS, I would reference this as SURVEY.UTDRIVERS.

If no libref is mentioned, SAS assumes that your dataset is in the work (temporary) library. If my previously used example of a dataset, called UTDRIVERS, was actually stored in the work file, I would reference it only as UTDRIVERS. Although the work library does not require a libref, WORK.UTDRIVERS and UTDRIVERS reference the same dataset.

Looking at Your Data

There are several ways to look at your data in SAS, each having its place depending on the purpose for looking at the data.

Spreadsheet View
A good way to look at your data is to double-click on the dataset to open up a spreadsheet view of the data as follows:

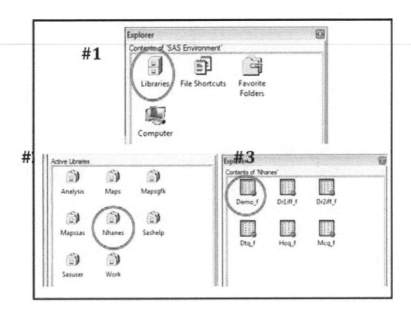

Figure 1.22

This method of looking at data results in a "viewtable," as shown in the following image:

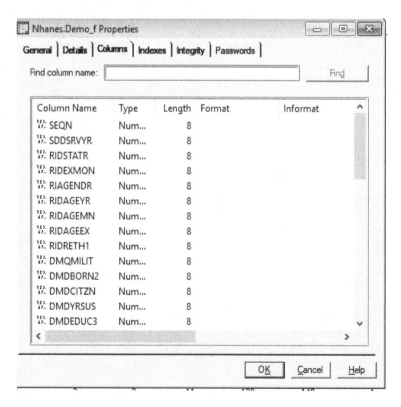

VIEWTABLE: Nhanes.Demo_f

	Respondent sequence number	Data Release Number	Interview/Examination Status	Six month time period	Gender
1		6	2	1	1
2		6	2	2	1
3		6	2	1	1
4		6	2	1	1
5		6	2	2	2
6		6	2	1	1
7		6	2	1	2
8		6	2	1	2
9		6	2	1	1

Figure 1.23

This is similar to looking at data in any other spreadsheet or database program. While there may be times that this is useful, it quickly becomes cumbersome and difficult if your dataset is large.

View Columns

A better way to look at your data would be to navigate through the explorer window to the dataset you want to open (as shown above), right-click on the dataset, and choose "View Columns." This will provide you with a list of variables and other summary information about the dataset, as shown in Figure 1.24:

Nhanes.Demo_f Properties

General | Details | Columns | Indexes | Integrity | Passwords |

Find column name: _____   Find

Column Name	Type	Length	Format	Informat
SEQN	Num...	8		
SDDSRVYR	Num...	8		
RIDSTATR	Num...	8		
RIDEXMON	Num...	8		
RIAGENDR	Num...	8		
RIDAGEYR	Num...	8		
RIDAGEMN	Num...	8		
RIDAGEEX	Num...	8		
RIDRETH1	Num...	8		
DMQMILIT	Num...	8		
DMDBORN2	Num...	8		
DMDCITZN	Num...	8		
DMDYRSUS	Num...	8		
DMDEDUC3	Num...	8		

OK   Cancel   Help

Figure 1.24

The Contents Procedure

The best way to look at your data is by using the contents procedure (step command PROC CONTENTS). PROC CONTENTS quickly outputs summary data on your dataset as well as a variety of information about the variables.

The syntax for PROC CONTENTS looks like this:

```
proc contents data=libref.datasetname;
run;
```

The libref = library name and datasetname = the name of the dataset you are looking at. The contents procedure output will appear in a results window and includes information such as the number of observations, the number of variables, the type of variables (numeric or character), formats (if any), and labels (if any). We will learn about formats and labels in a later lesson. If your dataset does not contain formats and /or labels, these columns will not appear in your PROC CONTENTS output.

The SAS System

The CONTENTS Procedure

Data Set Name	NHANES.DEMO_F	Observations	10537
Member Type	DATA	Variables	43
Engine	V9	Indexes	0
Created	12/11/2017 11:28:25	Observation Length	344
Last Modified	12/11/2017 11:28:25	Deleted Observations	0
Protection		Compressed	NO
Data Set Type		Sorted	YES
Label			
Data Representation	WINDOWS_64		
Encoding	wlatin1 Western (Windows)		

Figure 1.25   The first section of PROC CONTENTS gives you summary information, including the dataset name (notice the libref.datasetname), the number of observations, and the number of variables.

The default option in PROC CONTENTS is to list the variables in alphabetical order, as shown in Figure 1.26:

Alphabetic List of Variables and Attributes

#	Variable	Type	Len	Format	Informat	Label
187	Agec	Num	8			
174	BarEnScrn	Num	8	SCRN.		Barium Enema Screening Status
170	COL_BAR	Num	8			
183	CRC	Num	8	YN.		Have you ever been told by a health
180	Caprev	Num	8	NDK.		Ever had a Mammogram or PSA
12	CaseID	Num	8	11.	11.	CaseID

Figure 1.26

The columns from PROC CONTENTS include the following:

Type—NUM for numeric and CHAR for character variables.

Format—indicates if there are formats associated with any of the variables.

Label—indicates if there are labels associated with any of the variables.

(If your dataset does not contain formats or labels, these columns will not appear.)

Note 1.3 ▼

Numeric variables use numerals to define their values, and character variables use letters. Character variables are very sensitive to capitalization, spacing, etc., making them more difficult to use. Most public health data are coded in numeric form, such as 1 = Male, 2 = Female. SAS only retains the 1 and 2 in the database, and formats are used to describe them as male and female.

SAS allows you to choose an alternate display, if you desire.

- Using the option POSITION will give you a variable list in creation order or the order that they appear in the data spreadsheet from left to right.
- Using the option SHORT will list only the variable names in a row-by-row format.

The syntax used to list the variables in creation order looks like this:

```
proc contents data=libref.datasetname position;
run;
```

Which results in the following table:

Variables in Creation Order					
#	Variable	Type	Len	Format	Informat	Label
1	ID	Num	8	11.	11.	ID
2	agegrp	Char	255	$255.	$255.	agegrp
3	HighAA	Num	8			HighAA
4	PATIENTID	Num	8	11.	11.	PATIENTID
5	DOB_Combine	Num	8	DATETIME20.	DATETIME20.	DOB_Combine

Figure 1.27

To list the variables using the SHORT option, use the following syntax:

```
proc contents data=libref.datasetname short;
run;
```

Which gives results such as the following:

Figure 1.28

Saving SAS Programs

Now that we've written our first pieces of code (libname and PROC CONTENTS), we need to know how to save our SAS program. It is always a good practice to save your SAS programs so that you can use them later or reference them as needed. Saving your SAS program is very similar to saving a file in most other windows-driven programs.

With your cursor in the editor window, go to the "File" menu in the top left of the SAS window. Select "Save As." A new window will pop up, and you can navigate to where you would like to save your file.

Opening Previously Saved SAS Programs

Once you have a saved a SAS program, you may wish to return to the file to continue your work. To open your saved SAS programs, you should first open the SAS program.

Once SAS opens, go to the File Menu, and select "Open Program."

Figure 1.29

Navigate to where you saved the program and double-click. Here you can see that I have two SAS programs saved. I'll open Program 1.sas.

Figure 1.30

Program 1 will now open, and I can continue where I left off.

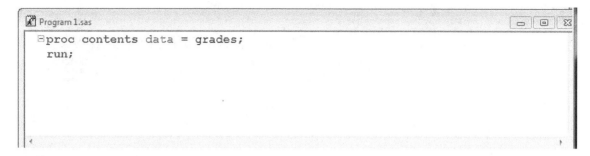

Figure 1.31

Summary

SAS has several windows and menus for performing analysis. It is important to become familiar with these tools and their functions to be able to use SAS effectively and efficiently.

Using SAS requires writing SAS code, which is made up of SAS statements that start with a SAS keyword or step command and end with a semicolon. Groups of SAS statements make up SAS steps, including DATA steps and Procedure (PROC) steps. Using these steps, you can manipulate and analyze your data.

One of the first steps you will need to do to work with data is to create a permanent library using a libname statement or the point-and-click method to help SAS locate your data. These permanent libraries allow you to save your work wherever your data are located. In addition, you may use the work library. The work library, however, empties each time you close SAS, and your data will not be saved.

Once SAS can find your data in the work or permanent libraries, you can use a variety of steps to look at your data, including opening the dataset using "View Columns" and PROC CONTENTS. Additional options allow you to customize viewing based on the information you need to obtain about your dataset.

Lesson #1—Study Questions

1. Name the three windows in SAS and explain how each is used.
2. List the two types of steps in SAS and explain how they are different.
3. Distinguish between SAS statements and SAS steps.
4. What is the most important punctuation in SAS?
5. Explain the structure of a SAS dataset.
6. What are the two methods you can use to create a permanent SAS library?
7. What is the name of the temporary library in SAS?
8. How does a temporary library differ from a permanent library?
9. If you put a dataset called SURVEYDATA in a permanent library called RESEARCH, how must you reference the dataset when you are working with it?
10. What is the order of the default variable listing from PROC CONTENTS?
11. Name two options in PROC CONTENTS and explain how they change the output.

12. Identify the syntax color for the following SAS syntax items:
    a. keywords
    b. step commands
    c. operators
    d. options
    e. comments
    f. variable names
    g. dataset names
    h. format names
    i. numbers/numerals
    j. titles
13. Explain how to save a SAS program.
14. Explain how to open a previously saved SAS program.

Lesson 2

Exploring Your Data

Learning Outcomes—Students Will:

- ▶ Demonstrate the use of a data dictionary to learn about a SAS dataset,

- ▶ Describe the types of "dirty data," and

- ▶ Use PROC PRINT to learn about a dataset.

Understanding Your Data

In the previous lesson, we used PROC CONTENTS to get summary information about the datasets and a list of variables included in the data. Beyond this, however, it is important to understand the data you are working with. Specifically, you need to understand where the data came from, how they are coded, and what any underlying codes mean. Many surveys, including NHANES, use skip patterns, which mean that not everyone who participates in the study will be asked each question. Understanding how these skip patterns influence your population of respondents is very important. Further, some survey respondents may not have the information requested or may refuse to answer some or all questions. A data dictionary can help you more closely examine your data.

Data Dictionaries

A data dictionary, also called a codebook or data documentation, is documentation about your dataset and the variables it contains. Most national datasets (as we are using in this book) come with data dictionaries created for them. Data dictionaries vary in format and detail; however, they typically include information about the variable name (i.e., what the column is labeled in the SAS spreadsheet), a description of the variables, the response options, and often the frequency of each response. Data dictionaries will also tell you who was eligible to be asked a question (also called the target or the universe). In many cases, not everyone who takes a survey is asked a question, because they are not eligible to be asked it. For example, questions about a spouse will not be asked of unmarried respondents, and questions about mammography screening will not be asked of male respondents or female respondents younger than the recommended age for screening mammography. When noneligible persons are not asked a question, this is called a skip pattern. Understanding the skip patterns in your data will ensure that the variables you use are answering the right questions. Links to the data dictionaries for all of the datasets used in the class have been provided to you on Learning Suite.

The following are two examples from the 2005–2006 NHANES Demographics DataSet data dictionary for variables DMDBORN and RIDAGEYR.

DMDBORN:

	DMDBORN		3	Target	
1				B(0 Yrs. to 150 Yrs.)	
	Hard Edits		4	SAS Label	
				Country of Birth - Recode	
2	English Text: In what country {were you/was SP} born?				
	English Instructions:				
5	Code or Value	6 Description	7 Count	Cumulative	Skip to Item
	1	Born in 50 US States or Washington. DC	8794	8794	8
	2	Born in Mexico	961	9755	
	3	Born Elsewhere	588	10343	
	7	Refused	5	10348	
	.	Missing	0	10348	

Figure 2.1

From this entry in the data dictionary, we learn the following:

1. The variable name: DMDBORN.
2. The question: "In what country {were you/was SP} born?"

Note 2.1 ▼

SP = Survey Participant. This is used when the questions were asked of a proxy, as is the case for children or persons physically or mentally unable to be interviewed directly.

1. The target: Anyone from 0 to 150 years of age (the entire study population).
2. The SAS label: In this example, it is "Country of Birth-Recode," which corresponds to what we see in our PROC CONTENTS:

11	DMDBORN2	Num	8	BEST12.	F12.		Country of Birth - Recode

Figure 2.2

3. The codes for respondents' answers: In this example, we can determine that the variable DMDBORN is categorical and has three valid options (1, 2, 3), one option for refused (7), and a missing designation (.).
4. The variable description: This gives the answer the respondent chose or a category into which that answer was placed (Born in the 50 US States or Washington, DC. Born in Mexico, etc.).
5. The count: The frequency of that response is provided.
6. Skip pattern instructions: A column is also provided for any skip pattern instructions. It's important to pay attention to these skip patterns, as they may explain some of the response variations, especially what may have been viewed as missing responses.

RIDAGEYR:

RIDAGEYR		Target		
		B(0 Yrs. to 150 Yrs.)		
Hard Edits		**SAS Label**		
		Age at Screening Adjudicated - Recode		
English Text: Best age in years of the sample person at time of HH screening. Individuals 85 and over are topcoded at 85 years of age.				
English Instructions:				
Code or Value	**Description**	**Count**	**Cumulative**	**Skip to Item**
0 to 84	Range of Values	10178	10178	
85	>= 85 years of age	170	10348	
.	Missing	0	10348	

Figure 2.3

From the entry in the data dictionary for variable RIDAGEYR, we can follow through all of the same content, as in the previous example. There is, however, one difference: RIDAGEYR is a continuous variable. We know this because instead of giving a set of valid values, the data dictionary provides a range of valid values. In this particular case, RIDAGEYR is only approximately continuous, as the data dictionary gives a range of valid values from 0 to 84 years and shows 85 as the value for anyone 85 or older (this is called topcoding and is done to preserve the anonymity of survey participants). The range of valid values from 0 to 84 indicates that observations that are 0, 1, 2, 3, 4, 5, 6, 7, 8, 9 ... 81, 82, 83, 84 are all valid (and likely present in the data).

"Dirty Data"

Part of understanding your data involves knowing which data are "dirty data." "Dirty data" refers to data that are missing, have invalid values (refused/don't know, etc.), are skipped either because of oversight or because of a skip pattern, are out of range, have an illegal value, or are incompatible with the other information given for the respondent.

Invalid Values

We include don't know and refused as "invalid values," because in most (but not all) situations, a response of don't know or refused does not allow you to infer anything about the respondent. There are few exceptions to this rule, but generally, responses of don't know and refused should be converted to missing and omitted from the analysis.

Out of Range Values

Out of range values are data values that are not possible in your dataset. For example, an age <0 or 140 years would not be possible. Illegal values might include a birth date in the future or a date out of range for the month (e.g., 30 days in February).

Incompatible Data

Incompatible data includes having a response for a question that should have been skipped because of a skip pattern (e.g., a man answering a question about mammography screening) or data that do not make sense when compared to other variables (e.g., a birth weight of 7 pounds and gestational age at birth of 24 weeks; age 16 years and Medicare beneficiary).

It is important to identify this "dirty" data to "clean" the data. Data cleaning will be discussed in a later lesson.

Skip Patterns in Data Dictionaries

Let's look at an example of how skip patterns might appear in a data dictionary. This exercise should illustrate the importance of looking closely at your data.

These two variables, KIQ022 and KIQ025, are from the Kidney Conditions-Urology (KIQ_U_D) NHANES dataset. Let's first notice that the sample size for this dataset is significantly smaller than the demographics data we were working with earlier. For both variables, the cumulative frequency is 4,979 observations. This indicates that the Kidney Conditions survey module was only given to a subset of NHANES respondents. This is fairly typical of large public health surveys. To minimize respondent burden, certain specialty topics are only asked of a subset of respondents.

The first variable in this sequence is KIQ022 "Have you ever been told by a doctor or health care professional that you had weak or failing kidneys?" Notice that people who answer "No" (coded as 2), or who refused to answer, or did not know the answer (coded as 7 or 9) skip to question KIQ005, thus skipping over and not being asked question KIQ025 "In the past month, have you received dialysis?"

The skip pattern is apparent if you are looking at the entire series of variables, but let's assume you are only looking at KIQ025.

KIQ022 - Ever told you had weak/failing kidneys

Variable Name: KIQ022

SAS Label: Ever told you had weak/failing kidneys

English Text: {Have you/Has SP} ever been told by a doctor or other health professional that {you/s/he} had weak or failing kidneys? Do not include kidney stones, bladder infections, or incontinence.

Target: Both males and females 20 YEARS - 150 YEARS

Code or Value	Value Description	Count	Cumulative	Skip to Item
1	Yes	136	136	
2	No	4829	4965	KIQ005
7	Refused	0	4965	KIQ005
9	Don't know	14	4979	KIQ005
.	Missing	0	4979	

KIQ025 - Received dialysis in past 12 months

Variable Name: KIQ025

SAS Label: Received dialysis in past 12 months

English Text: In the past 12 months, {have you/has SP} received dialysis (either hemodialysis or peritoneal dialysis)?

Target: Both males and females 20 YEARS - 150 YEARS

Code or Value	Value Description	Count	Cumulative	Skip to Item
1	Yes	17	17	
2	No	119	136	
7	Refused	0	136	
9	Don't know	0	136	
.	Missing	4843	4979	

KIQ005 - How often have urinary leakage

Variable Name: KIQ005

SAS Label: How often have urinary leakage

English Text: Many people have leakage of urine. The next few questions ask about urine leakage. How often {do you/does SP} have urinary leakage? Would {you/s/he} say . . .

English Instructions: CAPI INSTRUCTION: HELP SCREEN: Other terms for urinary

Figure 2.4

Looking only at this entry in the data dictionary, there is no obvious reference to the skip pattern. Examining the "target" does not explain this large number of missing, as the target appears to be inclusive of most adults in the survey. One clue that a skip pattern is at work is the number of missing observations. Here the number of missing is 4,843, a full 97% of the persons responding to the Kidney Conditions module. When large numbers of missing observations are present, review the variable or sometimes several variables preceding this index variable to ensure that you understand the data. Here, going back to the previous variable KIQ022, we easily identify the skip pattern, account for the large number of missing, and therefore correctly calculate data.

KIQ025 - Received dialysis in past 12 months

Variable Name:	KIQ025
SAS Label:	Received dialysis in past 12 months
English Text:	In the past 12 months, {have you/has SP} received dialysis (either hemodialysis or peritoneal dialysis)?
Target:	Both males and females 20 YEARS - 150 YEARS

Code or Value	Value Description	Count	Cumulative	Skip to Item
1	Yes	17	17	
2	No	119	136	
7	Refused	0	136	
9	Don't know	0	136	
.	Missing	4843	4979	

Figure 2.5

KIQ022 - Ever told you had weak/failing kidneys

Variable Name:	KIQ022
SAS Label:	Ever told you had weak/failing kidneys
English Text:	{Have you/Has SP} ever been told by a doctor or other health professional that {you/s/he} had weak or failing kidneys? Do not include kidney stones, bladder infections, or incontinence.
Target:	Both males and females 20 YEARS - 150 YEARS

Code or Value	Value Description	Count	Cumulative	Skip to Item
1	Yes	136	136	
2	No	4829	4965	KIQ005
7	Refused	0	4965	KIQ005
9	Don't know	14	4979	KIQ005
.	Missing	0	4979	

Figure 2.6

Let's further examine a scenario in which failing to understand the skip pattern could lead to incorrect conclusions. Let's suppose that I wanted to know the prevalence of US adults who were receiving dialysis treatment. I searched the codebook for the word dialysis, and the variable KIQ025 pops up. I quickly write code in my SAS program (which we will cover in the next chapter), resulting in the following table.

The FREQ Procedure

Received dialysis in past 12 months

KIQ025	Frequency	Percent	Cumulative Frequency	Cumulative Percent
1	17	12.50	17	12.50
2	119	87.50	136	100.00

Frequency Missing = 4843

Figure 2.7

I, therefore, conclude that 12.5% of the US population 20 and older has received dialysis treatment in the past year. If I do this, I will overestimate the prevalence of dialysis by about 12%, a significant overestimation. Considering both variables together, I determine that prevalence of dialysis is approximately 0.3% (17/4,979).

The 12.5% calculated as 17/136 represents the percentage of those who have been told they had weak or failing kidneys who have received dialysis. This underscores the importance of understanding skip patterns and the denominator you are working with in any measure of disease frequency.

Creating Your Own Data Dictionaries

As we progress through the class, we'll begin creating and modifying the variables in the NHANES dataset. It is always good practice to create data dictionaries for new variables you create or any variables you modify. A WORD table is an easy way to create a useful data dictionary for recoded variables. The following is a partial data table from a study I conducted that illustrates one method for putting together a data dictionary. The table can be expanded to include as many rows as necessary. In future lessons, you will be asked to create data dictionaries for recoded, new, or modified variables.

Table 2.1

Variable Name	Variable Label	Values	Format	Recoded From:	Comments
Gender	Gender	1 = Male	Gender	Q1	
Age	Age in years (continuous)	21–44 years	None	Q2	
Father	Father?	1 = Yes 2 = No	Ynf	Q3	

Using SAS Procedures to Look at Your Data

In combination with the data dictionary and PROC CONTENTS, the print, frequency, means, and univariate procedures can be useful in helping you check your work and in identifying "dirty" data. In this lesson, we'll cover the PROC PRINT. In Lesson 3, we will cover PROC MEANS and PROC FREQ.

The Print Procedure (PROC PRINT)

The print procedure allows you to display portions of your SAS dataset directly on the screen. The print procedure can be very useful for understanding the data and for checking your work. However, it is not useful for looking at large datasets.

The syntax for the print procedure is as follows:

```
proc print data=libref.datasetname;
var variable1 variable2;
run;
```

Where libref and datasetname refer to the compound (two-part) name of the dataset you would like to print data from. VAR is the SAS keyword that tells SAS which variables you would like to print. Here I have listed two variables; you may run PROC PRINT on as many variables as you desire (with variable names separated by a space), although it is recommended that you restrict use of this procedure to a small number of variables, as the resulting output can quickly become burdensome. As before, you will see that this particular SAS procedure consists of three SAS statements, each beginning with a SAS keyword or step command and ending with a semicolon.

```
1  proc print data=libref.datasetname;
2  var variable1 variable2;
3  run;
```

Figure 2.8

For the following examples, I will continue to use variables DMDBORN and RIDAGEYR from the demographics dataset. Using the following code, where my libref=nhanes, my dataset=demo_d, and my variables are DMDBORN and RIDAGEYR, I can print out the observations:

```
proc print data=nhanes.demo_d;
var dmdborn ridageyr;
run;
```

This code produces the following output (partial):

The SAS System

Obs	DMDBORN	RIDAGEYR
1	1	0
2	1	11
3	1	15
4	1	85
5	1	44

Figure 2.9 In this case, I can see that for the first observation (row) in the dataset, the respondent has a value of 1 for DMDBORN and 0 for RIDAGEYR. The complete output would print out one row for each person in the dataset, in this case over 10,000.

Print Procedure Options

The print procedure has a few useful options, including NOOBS, LABEL, and (OBS=xx). In PROC PRINT, the options are listed following the dataset name. As they are SAS keywords, when used correctly, they should display in light blue. Although these options are demonstrated only one at a time, the options may be combined to produce the output you desire.

NOOBS

Option NOOBS suppresses the column with the observation number. The syntax is as follows:

```
proc print data=nhanes.demo_d noobs;
var dmdborn ridageyr;
run;
```

Using this code produces this output (partial):

The SAS System

DMDBORN	RIDAGEYR
1	0
1	11
1	15
1	85
1	44
1	70

Figure 2.10

As you can see, the output is the same, except that the column labeled OBS (observations) is no longer printed. This may be useful if you are printing a large number of variables, as it will reduce the width of your table.

LABEL

The option for LABEL changes the column header from the variable name to the variable label (a more descriptive header). The code to include labels is as follows:

```
proc print data=nhanes.demo_d label;
var dmdborn ridageyr;
run;
```

This code gives the following output (partial):

Obs	Country of Birth - Recode	Age at Screening Adjudicated - Recode
	The SAS System	
1	1	0
2	1	11
3	1	15
4	1	85

Figure 2.11 Comparing the previous output, you can see that instead of labeling the columns as DMDBORN and RIDAGEYR, they are now listed as Country of Birth – Recode and Age at Screening Adjudicated – Recode. These labels are the same as those listed in the data dictionary and in PROC CONTENTS.

(OBS=xx)

The final option we will discuss for PROC PRINT is perhaps the most useful. The option (OBS=xx) is used to restrict the number of observations printed out by the print procedure. The xx should be replaced with whatever number of observations you wish to produce. For example, if I want to print only the first 15 observations, I could do so using the following code:

```
proc print data=nhanes.demo_d (obs=15);
var dmdborn ridageyr;
run;
```

Note that (OBS=15) is the correct syntax for this option. The parentheses must be included.

The SAS System

Obs	DMDBORN	RIDAGEYR
1	1	0
2	1	11
3	1	15
4	1	85
5	1	44
6	1	70
7	1	16
8	1	73
9	1	0
10	1	41
11	1	14
12	1	3
13	1	18
14	1	11
15	1	16

Figure 2.12   Here you can see that only the first 15 observations are output. This is particularly useful for checking your work when creating a variable that is a mathematical function of another.

Summary

Beyond getting summary information about datasets and their variables (as discussed in Chapter 1), it is important to understand the data you are working with, specifically, where the data came from, how they are coded, and what the underlying codes mean. A data dictionary or codebook can help you examine your data more closely by typically describing information about the variable name, a description of the variables, the response options, the frequency of responses, and the eligibility of a survey participant (SP) to respond to a question.

"Dirty data" refers to data that are missing, have invalid values (refused, don't know, etc.), are skipped because of oversight or because of a skip pattern, are out of range, have an illegal value, or are incompatible with the other information given for the respondent. It is especially important to recognize skip patterns to account for large numbers of SPs coded as missing and, therefore, correctly calculate data.

The print procedure (PROC PRINT) allows you to display portions of your SAS dataset directly onto your screen. While not useful for looking at large datasets, the PROC PRINT command will be key in helping you understand the data and in checking your work.

1. What is a data dictionary, and what types of information does it contain?
2. List two forms of data that would be considered "dirty data."
3. Explain what a skip pattern is and how it might be denoted in a data dictionary.
4. What is the purpose of using the print procedure?
5. What do the options OBS, NOOBS, and LABEL do when used with the print procedure?

Lesson 3

PROC MEANS, FREQ, and UNIVARIATE

Learning Outcomes—Students Will:

- ▶ Demonstrate use of the frequency procedure to determine frequency and proportions,
- ▶ Use the means procedure to generate descriptive data on continuous variables,
- ▶ Use the univariate procedure to assist in data cleaning,
- ▶ Save SAS output in HTML files, and
- ▶ Save SAS output using the ODS system to generate RTF and PDF files.

Using SAS Procedures to Look at Your Data

In combination with the data dictionary, PROC CONTENTS, and PROC PRINT that we have previously discussed, the frequency, means, and univariate procedures can be useful in helping you examine recoded variables to verify your work, identify "dirty" data, and produce useful statistics, such as frequencies, proportions, means, and standard deviations. In this lesson, we will cover PROC FREQ, PROC MEANS, and PROC UNIVARIATE.

The Frequency Procedure (PROC FREQ)

The frequency procedure (PROC FREQ) is useful for looking at categorical data. PROC FREQ produces a count of how many records (observations or rows) in your dataset have each value for the variable or variables you request output on. The count is also called a frequency and is where the frequency procedure gets its name. In addition to the frequency, the default output also produces a percentage, cumulative frequency, and cumulative percentage. In later lessons, we'll learn how the frequency procedure can be used to produce other types of percentages as well as several statistics and measures of association.

The syntax for the frequency procedures follows this pattern:

```
proc freq data=datasetname;
tables variablename;
run;
```

The name of any variable or variables contained in the dataset can be listed following the TABLES command. As before, variables should be separated with a space. Note that the SAS keyword here is TABLE or TABLES (SAS doesn't care) in contrast to the other procedures we have learned where the keyword is VAR.

The default output will produce one table for each variable in the TABLES statement that looks like this:

				The FREQ Procedure		
		Marital Status		
DMDMARTL	Frequency	Percent	Cumulative Frequency	Cumulative Percent
1	2711	40.47	2711	40.47
2	463	6.91	3174	47.39
3	471	7.03	3645	54.42
4	164	2.45	3809	56.87
5	2412	36.01	6221	92.88
6	468	6.99	6689	99.87
77	8	0.12	6697	99.99
99	1	0.01	6698	100.00
Frequency Missing = 3650				

Figure 3.1   This table is the frequency procedure output for the variable DMDMARTL from the NHANES 2005–2006 demographics dataset. The values listed here are 1, 2, 3, 4, 5, 6, 77, and 99. From the data dictionary (shown next), we confirm that 77 and 99 are the "refused" and "don't know" values, while values 1–6 are valid responses. We also see that there are 3,650 observations that are missing.

Looking at the data dictionary, we see that this question about marital status was only asked of respondents 14 years or older. This explains the large number of missing. Remember that large numbers of missing may indicate a limited universe (who is asked the question) or a skip pattern, so always check to see why observations are missing, as this will affect your use of the variables.

DMDMARTL	Target			
	B(14 Yrs. to 150 Yrs.)			
Hard Edits	SAS Label			
	Marital Status			
English Text: Marital Status				
English Instructions:				
Code or Value	Description	Count	Cumulative	Skip to Item
1	Married	2711	2711	
2	Widowed	463	3174	
3	Divorced	471	3645	
4	Separated	164	3809	
5	Never married	2412	6221	
6	Living with partner	468	6689	
77	Refused	8	6697	
99	Don't know	1	6698	
.	Missing	3650	10348	

Figure 3.2

Frequency Procedure Options

As with the other procedures, the frequency procedure also has a variety of options. At this point, we will learn only two: /MISSING and /MISSPRINT. Both of these options control how missing values are displayed in the frequency procedure output. As noted earlier in the default output, missing values are listed as a footnote below the table. Both /MISSING and /MISSPRINT list the missing values in the body of the table but differ in whether the missing values are included in percentages. To use /MISSING or /MISSPRINT, list either option in the TABLES statement, following the variable list and prior to the semicolon. To separate the variable names from the options, the forward slash (/) is used.

/MISSING
/MISSING tells SAS to include the missing values in the frequency table, including the missing values in the percentages.
   The syntax is as follows:

```
proc freq data=libref.datasetname;
tables variablename /missing;
run;
```

Using the earlier example of variable DMDMARTL (Marital Status), the following table is produced:

	The FREQ Procedure			
	Marital Status			
DMDMARTL	Frequency	Percent	Cumulative Frequency	Cumulative Percent
.	3650	35.27	3650	35.27
1	2711	26.20	6361	61.47
2	463	4.47	6824	65.95
3	471	4.55	7295	70.50
4	164	1.58	7459	72.08
5	2412	23.31	9871	95.39
6	468	4.52	10339	99.91
77	8	0.08	10347	99.99
99	1	0.01	10348	100.00

Figure 3.3

As you can see, the missing values are now included as the first row in the table. You can see that this also includes the missing values as part of the percentages. This has the advantage of telling me what percentage of the total observations are missing (35.27%) but is not useful in most analyses where missing data should be excluded. **The /MISSING option should always be used in data management tasks such as checking recodes, but it should not be used in analysis.**

/MISSPRINT

A second option /MISSPRINT includes the missing observations in the table but not in the percentages. As before, the option is included, following the TABLES statement.

The syntax is shown next:

```
proc freq data=nhanes.demo_d;
tables dmdmartl /missprint;
run;
```

This produces the following output:

DMDMARTL	Frequency	Percent	Cumulative Frequency	Cumulative Percent
.	3650	.	.	.
1	2711	40.47	2711	40.47
2	463	6.91	3174	47.39
3	471	7.03	3645	54.42
4	164	2.45	3809	56.87
5	2412	36.01	6221	92.88
6	468	6.99	6689	99.87
77	8	0.12	6697	99.99
99	1	0.01	6698	100.00

The FREQ Procedure

Marital Status

Frequency Missing = 3650

Figure 3.4

Note that although the missing values are still listed in the table, they are not included in the percentages. From the prior table where missing values are included in the percentage, we would conclude that 26.20% of the sample was married (value=1), whereas in this table, we would conclude 40.47% of the sample was married. Very different answers!

Multiple Variables and Multiple Tables

As with the other procedures we have learned, you can include multiple variables in a single TABLES statement within PROC FREQ, as shown here:

```
proc freq data=nhanes.demo_d;
tables dmdmartl dmdborn;
run;
```

You may also include multiple TABLES statements in a single PROC FREQ. For example, this syntax will produce a total of four tables: one for DMDMARTL with the missing on the table and one without, and one for DMDBORN with missing on the table and one without:

```
proc freq data=nhanes.demo_d;
tables dmdmartl dmdborn;
tables dmdmartl dmdborn /missing;
run;
```

Cross Tabulating with PROC FREQ

The frequency procedure can also be used to cross tabulate two variables. Cross tabulation is looking at one variable by the levels of another. The classic epidemiologic cross tabulation is the 2 x 2 or contingency table. To cross tabulate two variables, list them in the TABLES statement separated by an asterisk. SAS produces "row by column" (rxc) tables so that the first variable in the TABLES statement is your row variable and the second is your column variable, as follows:

```
proc freq data=libref.datasetname;
tables var1*var2;
run;
```

The cross tabulation is useful for many things, but particularly for checking your recodes. There are a variety of statistics that can be calculated from the cross tabulation, but we'll learn about those later.

Here's an example of a cross tabulation where I'm looking at gender (RIAGENDR) and ethnicity (RIDETH1):

```
proc freq data=nhanes.demo_d;
tables riagendr*rideth1;
run;
```

Producing the following output:

The default output for a cross tabulation gives you four numbers:

1) The Frequency
2) The Total Percent
3) The Row Percent
4) The Column Percent

A legend is provided in the upper right corner.

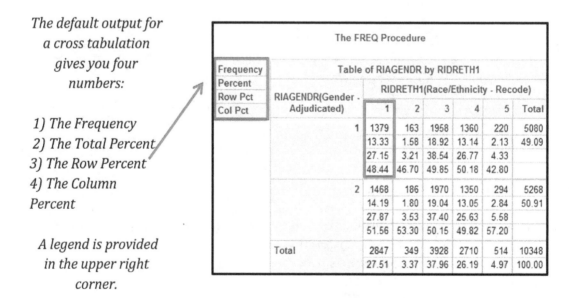

The FREQ Procedure

Frequency / Percent / Row Pct / Col Pct	RIAGENDR(Gender - Adjudicated)	RIDRETH1(Race/Ethnicity - Recode)					Total
		1	2	3	4	5	
	1	1379	163	1958	1360	220	5080
		13.33	1.58	18.92	13.14	2.13	49.09
		27.15	3.21	38.54	26.77	4.33	
		48.44	46.70	49.85	50.18	42.80	
	2	1468	186	1970	1350	294	5268
		14.19	1.80	19.04	13.05	2.84	50.91
		27.87	3.53	37.40	25.63	5.58	
		51.56	53.30	50.15	49.82	57.20	
	Total	2847	349	3928	2710	514	10348
		27.51	3.37	37.96	26.19	4.97	100.00

Figure 3.5

Note that the first variable in my TABLES statement is RIAGENDR (Gender), and this appears in the rows of my data and that RIDRETH1, my second variable, appears in the columns (remember rxc). From the data dictionary, I find that for RIAGENDR: 1 = Male and 2 = Female and for RIDRETH1: 1 = Mexican American, 2 = Other Hispanic, 3 = Non-Hispanic White, 4 = Non-Hispanic Black, and 5 = Other Race/ Multiracial.

The three types of percentages are calculated using different denominators from the table. Understanding how this works is important, as each answers a different question.

The **percent or total percent** is the first percentage in the box. If we look at the top-left box, which measures males and Mexican Americans, we can see that there 1,379 Mexican American males in the sample. We have a total of 10,348 respondents (see bottom corner box). The total percent is calculated as (1,379/10,348)*100 = 13.33%. This tells us that approximately 13% of the total sample (10, 348 respondents) are Mexican American males. Taking the total percentage from each cell in the tables gives you 100%.

The **row percent** is calculated by dividing the number in the cell by the total in the row, in this case 5,090 males. Thus the row percent is (1,379/5,080)*100 = 27.15%. This can be interpreted by saying that approximately 27% of the males in the sample were Mexican American. The row percents add up to 100% for each row of the table.

Finally, the **column percent** is calculated by dividing the cell value by the total observations in the column, in this case 2,847 Mexican Americans. The column percent is calculated as (1,379/2,847)*100 = 48.44%. Thus 48% of Mexican Americans in the sample are male. The column percents add up to 100% for each column in the table.

Because the total percent, column percent, and row percent all answer different questions, it is very important that you understand what the percentage means so that you can correctly interpret it. If in doubt, break out a calculator and manually calculate percentages until you are sure you understand how to read this table.

Controlling Cross-Tabulation Output

Sometimes it may be desirable to reduce the amount of output for a cross tabulation. This can be done by adding any combination of the options NOFREQ, NOPERCENT, NOCOL, or NOROW.

- NOFREQ suppresses the frequency.
- NOPERCENT suppresses the total percentage.
- NOCOL suppresses the column percentages.
- NOROW suppresses the row percentages.

As with /MISSING and /MISSPRINT, these options are added after the TABLES statement, as shown here:

```
proc freq data=nhanes.demo_d;
tables riagendr*rideth1 /nocol norow nopercent;
run;
```

Here I am requesting a table without the column, row, or total percentages so that only the frequencies will print, as follows:

Frequency	Table of RIAGENDR by RIDRETH1						
	RIAGENDR(Gender - Adjudicated)	RIDRETH1(Race/Ethnicity - Recode)					
		1	2	3	4	5	Total
	1	1379	163	1958	1360	220	5080
	2	1468	186	1970	1350	294	5268
Total		2847	349	3928	2710	514	10348

Figure 3.6

Reducing the output is useful for (at least) two reasons: (1) Restricting the output reduces the size of your table, which may make viewing the table or checking recodes easier. (2) If I restrict the output to only those counts or percentages I intend to report, I reduce the likelihood of transcription errors with populating result write-ups or tables.

You may also use the /MISSING or /MISSPRINT options with cross tabulations, as demonstrated next:

```
proc freq data=nhanes.demo_d;
tables riagendr*rideth1 /nocol norow nopercent missing;
run;
```

In this example, I am looking for a cross tabulation of marital status (DMDRARTL) and gender (RIAGENDR), suppressing all output except for the frequency and requesting the missing to print in the table and be included in the percentages.

This results in the following output:

Frequency	Table of DMDMARTL by RIAGENDR			
	DMDMARTL(Marital Status)	RIAGENDR(Gender - Adjudicated)		
		1	2	Total
	.	1847	1803	3650
	1	1390	1321	2711
	2	126	337	463
	3	205	266	471
	4	73	91	164
	5	1219	1193	2412
	6	216	252	468
	77	3	5	8
	99	1	0	1
Total		5080	5268	10348

Figure 3.7

The Means Procedure

The means procedure is useful for values that are numeric and continuous, or approximately continuous, such as age, weight, height, distance, etc. As the name implies, PROC MEANS produces the mean, or average, of the observation values. By default, PROC MEANS produces the following statistics:

- N (number of observations)
- Mean (average)
- Standard deviation (dispersion from the mean)
- Min (lowest Value)
- Max (highest Value)

The means procedure is useful for data management because the mean is very sensitive to extreme values. In many datasets, the invalid data (don't know and refused) are denoted as extreme values. For example, in the figure above, you can see that 77 and 99 are used to denote refused and missing. These extreme values can skew the mean. For instance, if I ran a mean on age, and the mean age in my sample was 93 years, this could indicate to me that I had a problem with my data, as 93 years of age is an unlikely average for the US population.

The syntax for the means procedure is as follows:

```
proc means data=libref.datasetname;
var variablename;
run;
```

Using the variable RIDAGEYR (age at interview), I can run a PROC MEANS to gather information about the variable:

```
proc means data=nhanes.demo_d;
var ridageyr;
run;
```

As you can see in the following image, each of the statistics in the previous list is printed by default. This example shows 10,348 observations with a mean of 28 years, a range of 0–85 years, and a standard deviation of 24.08:

The MEANS Procedure

Analysis Variable : RIDAGEYR Age at Screening Adjudicated - Recode				
N	Mean	Std Dev	Minimum	Maximum
10348	27.9957480	24.0822577	0	85.0000000

Figure 3.8

From the NHANES Medical Conditions DataSet, we can see how the mean may be skewed by dirty data. Looking at variable MCQ180B—"Age when told you had heart failure"—we see that those who reported they did not know how old they were when they were first told they had heart failure are recorded as 99999.

MCQ180B		Target			
		B(20 Yrs. to 150 Yrs.)			
Hard Edits		SAS Label			
		Age when told you had heart failure			
English Text: How old {were you/was SP} when {you were/s/he was} first told {you/s/he} . . .had congestive heart failure?					
English Instructions: ENTER AGE IN YEARS. SKIP TO MCQ160C.					
Code or Value	Description		Count	Cumulative	Skip to Item
0 to 84	Range of Values		169	169	MCQ160C
85	85 years or older		9	178	MCQ160C
77777	Refused		0	178	MCQ160C
99999	Don't know		2	180	MCQ160C
.	Missing		9642	9822	MCQ160C

Figure 3.9

Running a PROC MEANS on this variable produces the following output:

The MEANS Procedure

Analysis Variable : MCQ180B Age when told you had heart failure				
N	Mean	Std Dev	Minimum	Maximum
180	1168.95	10505.20	0	99999.00

Figure 3.10

This says that the average age of diagnosis for heart failure is 1,169 years—an illogical answer. If we remove those invalid values from the variable (a skill covered in Lesson 8), we find that the real mean age at diagnosis was 58.5 years—a much more logical possibility.

Means Procedure Options

The means procedure does have a few options that may be useful to know about. These options include SUM, MEDIAN, N, STD, MIN, MAX, MEAN, and MAXDEC. Each of these options, with the exception of MAXDEC, control what output PROC MEANS produces. All options are requested after the dataset name in the procedure statement. When requesting options in PROC MEANS, it will only output what you have requested. That means that if you request a MEDIAN, you will only get the median, not the median in addition to the default output.

For example:

```
proc means data=nhanes.demo_d median;
var ridageyr;
run;
```

This results in the following output:

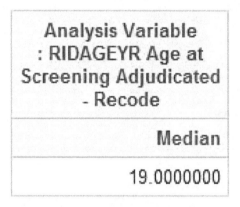

Figure 3.11

If I would like to view the median in addition to the default output, I need to request each item as follows:

```
proc means data=nhanes.demo_d n std min max mean median sum;
var ridageyr;
run;
```

Resulting in:

Analysis Variable : RIDAGEYR Age at Screening Adjudicated - Recode						
N	Std Dev	Minimum	Maximum	Mean	Median	Sum
10348	24.0822577	0	85.0000000	27.9957480	19.0000000	289700.00

Figure 3.12

The MAXDEC option is a very useful tool for modifying your output. This option tells SAS how many decimal places to print out. Typically, statistics should be reported out to only a few decimal places. By using the maxdec option, SAS does the work of rounding for you and reduces rounding errors. For example, requesting MAXDEC = 2 will bring all of the output within two decimal places, as you can see here:

```
proc means data=nhanes.demo_d maxdec=2;
var ridageyr;
run;
```

Analysis Variable : RIDAGEYR Age at Screening Adjudicated - Recode				
N	Mean	Std Dev	Minimum	Maximum
10348	28.00	24.08	0.00	85.00

Figure 3.13

The MAXDEC options can be combined with all other options in PROC MEANS.

The Univariate Procedure (PROC UNIVARIATE)

Much like PROC MEANS, the univariate procedure is used for looking at continuous variables. Like PROC MEANS, PROC UNIVARIATE gives you the N, mean, and the standard deviation. In addition to those basics, PROC UNIVARIATE also gives you the variance, standard error, median, mode, range and interquartile range, quantiles, and extreme observations. Extreme observations are particularly useful for identifying dirty data.

The syntax for PROC UNIVARIATE is as follows:

```
proc univariate data=libref.dataset;
var variable1 variable2;
run;
```

The following output gives the univariate statistics for RIDAGEYR:

The UNIVARIATE Procedure				Quantiles (Definition 5)	

The UNIVARIATE Procedure
Variable: RIDAGEYR (Age at Screening Adjudicated - Recode)

Moments			
N	10348	Sum Weights	10348
Mean	27.995748	Sum Observations	289700
Std Deviation	24.0822577	Variance	579.955138
Skewness	0.79979469	Kurtosis	-0.5054977
Uncorrected SS	14111164	Corrected SS	6000795.81
Coeff Variation	86.0211264	Std Error Mean	0.23673855

Basic Statistical Measures			
Location		Variability	
Mean	27.99575	Std Deviation	24.08226
Median	19.00000	Variance	579.95514
Mode	0.00000	Range	85.00000
		Interquartile Range	36.00000

Tests for Location: Mu0=0				
Test		Statistic	p Value	
Student's t	t	118.256	Pr > |t|	<.0001
Sign	M	4911	Pr >= |M|	<.0001
Signed Rank	S	24120377	Pr >= |S|	<.0001

Quantiles (Definition 5)	
Quantile	Estimate
100% Max	85
99%	85
95%	76
90%	67
75% Q3	45
50% Median	19
25% Q1	9
10%	2
5%	0
1%	0
0% Min	0

Extreme Observations			
Lowest		Highest	
Value	Obs	Value	Obs
0	10344	85	10158
0	10337	85	10198
0	10329	85	10279
0	10299	85	10289
0	10285	85	10335

Figure 3.14

Here you can see that the mean age is the same (28 years) as reported in the PROC MEANS, but you are also given a lot of other potentially useful information.

Controlling Output in SAS

If you've been following along in SAS, you may have noticed that your results window is getting crowded. In versions 9.3 or newer of SAS, the default output is HTML (all SAS output you have seen to this point in the workbook have been HTML). This output has the advantage of being easy to read and nice to look at. It is also easily copied and dropped into other types of files. You can clear this results window anytime you desire by typing in and running the following code:

```
ods html close;
ods html;
```

This gives you a clean slate but must be repeated each time you wish to start a new results window. Note that the window will not clear until you run the first PROC following these ODS commands.

One way to control the output produced is to type and run the following code at the beginning of each SAS program.

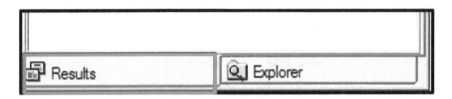

```
ods html newfile=proc;
```

Once executed, this code will open a new results window for each procedure that you run. To view previously run PROC steps, you can navigate through the results using the explorer window (choose the tab on the bottom of the explorer window that says "Results").

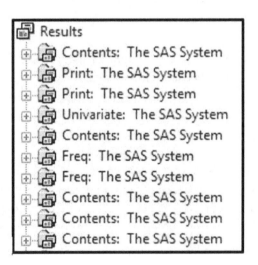

Figure 3.15

The results window contains a history of the PROC steps you have run during your SAS session. You can use the tree to navigate to the results you wish to view.

Figure 3.16

You can also view output using the listing option rather than HTML. Listing is the standard output in older versions of SAS. This is much less pretty, but the window can be easily cleared.

To turn on the listing option, complete the following:

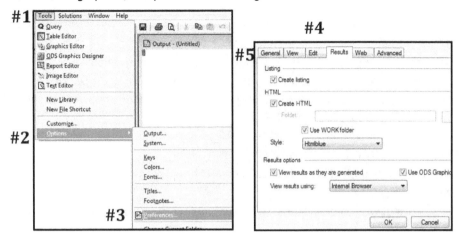

Figure 3.17

Now, looking at the results window, you can navigate to the listing and HTML views, depending on which one you need.

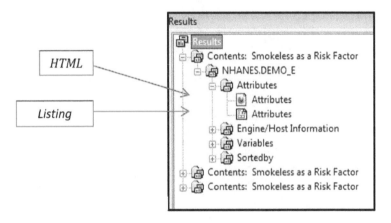

Figure 3.18

This is an example of listing output for DMDMARTL (Marital Status):

```
                      The SAS System

                     The FREQ Procedure

                      Marital Status

                                  Cumulative      Cumulative
DMDMARTL    Frequency    Percent   Frequency        Percent
       1         2711      40.47        2711          40.47
       2          463       6.91        3174          47.39
       3          471       7.03        3645          54.42
       4          164       2.45        3809          56.87
       5         2412      36.01        6221          92.88
       6          468       6.99        6689          99.87
      77            8       0.12        6697          99.99
      99            1       0.01        6698         100.00

              Frequency Missing = 3650
```

Figure 3.19

Although the information is the same, the formatting is not as nice. However, to clear the listing window, you only need to right-click and select "Clear All."

Saving SAS Output

There are many reasons you may wish to save your SAS output. Saving the output in an HTML, Rich Text File (RTF) or PDF allows you to open and look at output when you are away from SAS. This is particularly useful for populating tables or writing up results after the analysis is complete. Saving output may also enable you to share output with non-SAS users, RTF and PDF files are particularly useful. We will first discuss how to save HTML output, followed by an introduction to the SAS Output Delivery System (ODS) to create RTF and PDF files.

Saving HTML Files

All of the output we have looked at so far has been HTML output. This is by far the easiest type of output to save.

Here is my SAS window with one table in it. Often, this HTML output would contain many tables and results.

Figure 3.20

When I am in the results window, I can save this output by going to FILE—>SAVE AS and then navigating to the location I wish to save the file, similar to other windows-based programs. I have named my file "Dialysis Table" and saved it to a folder called "Output" on my flash drive.

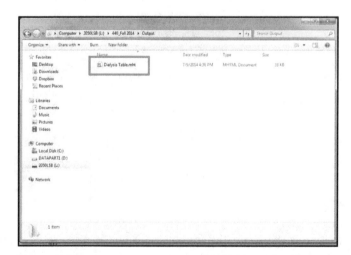

Figure 3.21

I can now open this file in any web browser. Double-clicking on this file will automatically open it in my default web browser here in Microsoft Explorer.

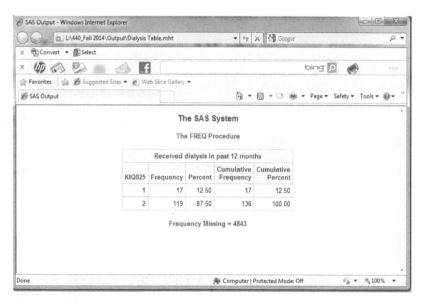

Figure 3.22

There are a couple of benefits to the HTML output. First, it is very easy to save, and second, it can be opened on any computer with a web browser. The major downside is that once saved, HTML output cannot be modified in anyway. Here is where saving output as an RTF file has real advantages.

Saving SAS Output as an RTF File

RTF files are a type of document file format that can be read by most word processing software. RTF files are more malleable than HTML output, allowing you to save entire projects worth of data into a single file and to add explanatory notes, titles, or commentary after the fact to customize the data to the audience you are sharing it with. RTF output is created using SAS's ODS and is requested by writing SAS syntax. The first line of syntax opens the RTF file, and the last line of syntax closes the RTF file. All of the code sandwiched between the two statements will be output to an RTF file when the statements are executed. You can have dozens of procedures between those two lines of code if desired. Note: the file will not be created until the ODS RTF CLOSE; statement is executed.

Let's look at the previous dialysis table as an example.

```
ods rtf file = 'pathway\documentname.rtf';
/*any procs you want to output*/
ods rtf close;
```

The pathway is the disk drive and tree of folders you wish to save the data in, and the document name is whatever you'd like to call the document followed by the file extension rtf. Here's an example producing the same table of variable KIQ025 we previously used.

```
ods rtf file = 'L:\440_Fall 2014\Output\Dialysis Table.rtf';
proc freq data = kiq_u_d;
tables kiq025;
run;
ods rtf close;
```

When I execute this code, the newly created RTF file will automatically open in the default word processor on my computer, in my case Microsoft Word.

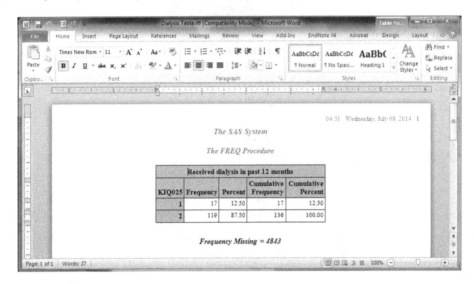

Figure 3.23

One major advantage to the RTF file is that I can now manipulate the output to add explanatory notes or titles, delete tables that are not relevant to the audience, etc. Here you can see that I've altered the title to be more descriptive:

Figure 3.24

RTF files are an incredibly useful output management tool. RTF files can be appended to reports or easily printed and filed for use later. Because most people have access to a word processor, this can allow for enhanced collaboration when working with others on research.

Saving SAS Output as a PDF File

Requesting output saved as a PDF file is similar in nearly all respects to an RTF file. The code is identical, except that you specify PDF instead of RTF. There are, however, two things to note. (1) When executing the code, the PDF will save but may not automatically open, depending on your computer settings. If the file doesn't automatically open, you'll need to go to the folder where you saved the file to open it manually. (2) PDF files are not as easily manipulated. Adobe Reader will allow you open these files, but you must have a version of Adobe Acrobat to make changes. This may also limit the ability of collaborators to easily comment on your output if they have questions or notes to make. The syntax for saving output as a PDF is as follows:

```
ods pdf file = 'pathway\documentname.pdf';
/*any procs you want to output*/
ods pdf close;
```

Note the only two changes are specifying pdf where I previously specified rtf. Here is the specific syntax for saving my dialysis tables as a pdf.

```
ods pdf file = 'L:\440_Fall 2014\Output\Dialysis Table.pdf';
proc freq data = kiq_u_d;
tables kiq025;
run;
ods pdf close;
```

The opened PDF file looks like this:

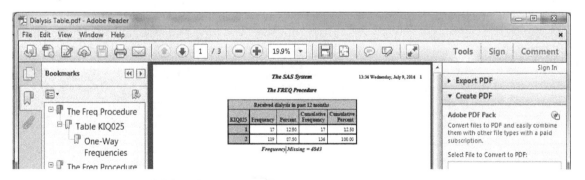

Figure 3.25

Perhaps one advantage of the PDF over RTF is that the PDF retains the results navigation, making it easier to navigate through large amounts of data.

Saving Output Summary

Each method of saving output has its advantages and disadvantages. For longer-term storage, I prefer the RTF or PDF options, particularly because they allow me to append the output tables to a written report or data dictionary or other documentation I might have associated with a single project. In general, it is not necessary to save all of your output from SAS, but you should certainly consider saving any output you may need to use again, including for results write-ups, populating tables, discussions with colleagues, or your own reference. Typically, after I have finished all my data management and worked all of the bugs out of my analytic approach, I will save the final output tables as an RTF or PDF file.

Lesson #3—Study Questions

1. What type of information can be obtained from running the default frequency procedure?
2. What is the difference between the options /missing and /missprint when used with the frequency procedure?
3. How do you cross tabulate two variables?
4. What types of additional information are provided when cross-tabulating data?
5. What order should variables be listed in the tables statement when cross tabulating?
6. List the four options for controlling cross-tabulation output.
7. List three types of information that can be obtained from running the means procedure.
8. What options are available to modify PROC MEANS output?
9. List three types of information that can be obtained from running the univariate procedure that cannot be obtained from running the default means procedure.
10. How does the syntax for PROC FREQ differ from PROC MEANS and PROC UNIVARIATE?
11. Discuss the three options for saving SAS output.

Lesson 4

PROC SGPLOT

Learning Outcomes—Students will:

> ▶ Use PROC SGPLOT to create descriptive graphs for categorical and continuous data.

In addition to reviewing data using numbers, the ability to visualize data with charts and graphs can be informative. Some characteristics of the data distribution and the relationships between two or more variables are clearer with visual presentations. This chapter will cover some basics of using the SGplot procedure. In particular, we will examine how to create histograms, scatterplots, bar charts, and box and whisker plots. Histograms, scatterplots, and boxplots are used to display continuous data. Bar charts are used to display categorical data.

About PROC SGPLOT

The SGPLOT procedure can create a total of 16 different types of graphs or charts. Additional "SG" procedures, such as SGPANEL, can be used to create more advanced graphics.
   The general syntax for PROC SGPLOT is as follows:

```
proc sgplot data = libref.datasetname;
keyword variablename;
run;
```

The procedures that we have looked at up to this point have had a single keyword to make them work (e.g., var, tables). Because PROC SGPLOT can create many different types of charts and graphs, the keyword changes here. "Keyword" in the aforementioned syntax is a placeholder that can be replaced with the keywords in the box to the right. Many keywords exist; we will focus on just a few of them in this chapter.

Making Histograms in PROC SGPLOT

Histograms show the distribution of continuous variables. Here is an example of creating a histogram with PROC SGPLOT for the variable BMXWT (weight). Notice that the keyword is HISTOGRAM. Keywords should turn blue to indicate that they are used with the appropriate procedure and spelled correctly.

```
proc sgplot data = combo;
histogram bmxwt;
run;
```

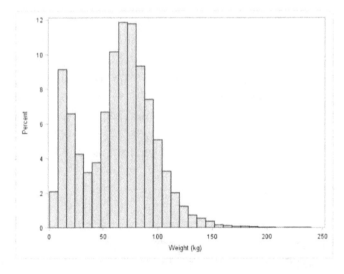

Figure 4 .1

You can also add a density curve to a histogram with a DENSITY statement. The density curve allows you to assess how well the variable conforms to a normal distribution. To add a density curve, you include both the HISTOGRAM and the DENSITY statement, with each followed by the variable you are graphing.

```
proc sgplot data = combo;
histogram bmxwt;
density bmxwt;
run;
```

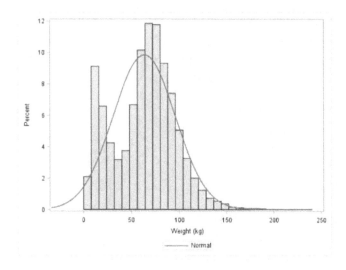

Figure 4 .2

Making Box and Whisker Plots in PROC SGPLOT

Box and whisker plots display the distribution of a single continuous variable. The keyword VBOX produces a vertical boxplot, while the keyword HBOX produces a horizontal boxplot. Choosing between vertical and horizontal plots is a personal preference. By default, SAS will plot the mean, median, interquartile range, minimum, and maximum, as well as some outlying data in the boxplot.

```
proc sgplot data=nhanes.bmx_f;
vbox bmxwt;
run;
```

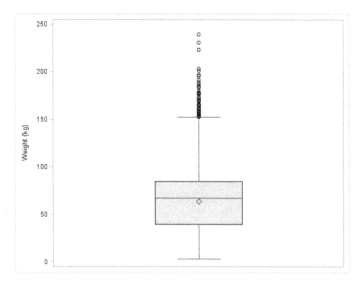

Figure 4 .3

Box and whisker plots can also be used to display the distribution of a continuous variable within the levels or strata of another variable. For example, you might expect that the distribution of weight (bmxwt) is different for males and females (riagendr). We can look at the data to examine this by using the CATEGORY option in the VBOX or HBOX statement.

```
proc sgplot data=combo;
hbox bmxwt /category = riagendr;
run;
```

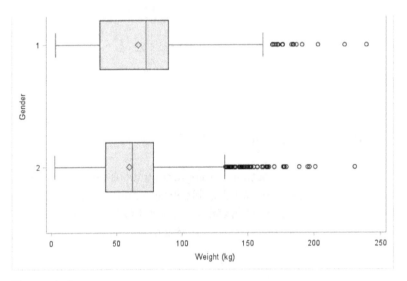

Figure 4 .4

Notice that SAS produces two different boxplots. The upper plot in this graph is labeled for gender = 1 and the lower graph is labeled for gender = 2. We can refer back to our data dictionary for assistance in identifying which gender value represents males and which represents females.

Making Scatterplots in PROC SGPLOT

Scatterplots are used for examining the distribution of two continuous variables. One of the variables is plotted on the x-axis and the other on the y-axis. The keyword SCATTER is used to create a scatterplot, and we specify which variable to put on the x-axis and which to put on the y-axis. Here we are looking at the association of weight (bmxwt) and height (bmxht).

```
proc sgplot data = combo;
scatter x=bmxwt y=bmxht;
run;
```

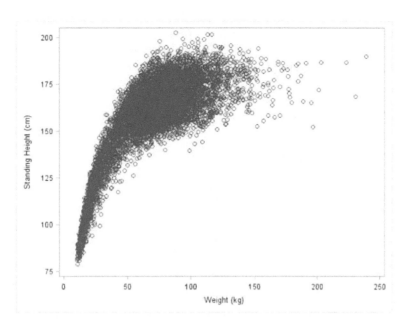

Figure 4 .5

Making Bar Charts in PROC SGPLOT

Bar charts are commonly used to chart the distribution of categorical variables. The keywords VBAR or HBAR are used to specify whether the bars should run vertically or horizontally. This vertical bar chart displays the distribution of marital status. The bars represent the frequency or count of people in each category.

```
proc sgplot data = combo;
vbar dmdmartl;
run;
```

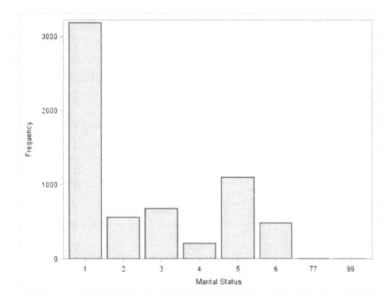

Figure 4 .6

Although the default presentation is to display the count or frequency in the bars, a more useful option is to display the percentage. Percentages are generally more interpretable and meaningful than counts. To change the statistics displayed, the option STAT is entered after a forward slash in the HBAR or VBAR statement.

```
proc sgplot data = combo;
hbar dmdmartl/ stat=percent;
run;
```

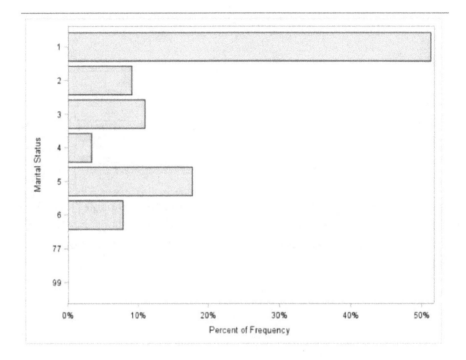

Figure 4 .7

Stacked bar charts are another useful variation on the bar chart. Stacked bar charts display the distribution of a categorical variable grouped by the strata of another variable. To add strata, the GROUP statement is added to the syntax as shown next. Notice that we continue to use the STAT = percent option to request the percentage. This chart shows the distribution of marital status by gender.

```
proc sgplot data = combo;
hbar dmdmartl/ stat=percent
group = riagendr;
run;
```

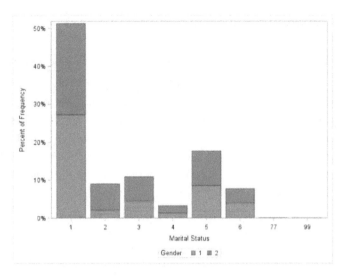

Figure 4 .8

Making Pie Charts in PROC GCHART

Pie charts are used to represent data as parts of a whole—for example, the percentage of a population with different education levels. Pie charts are prone to distortion and as such may not be the best option for displaying your data. In addition, pie charts should be used for a relatively small number of categories. If the pie needs to be divided into more than five to eight slices, it can be difficult to read.

As of the writing of this book, PROC SGPLOT does not have an option for creating a pie chart. However, a separate, older procedure called PROC GCHART can be used to create a pie chart. The syntax is similar to the SGPLOT syntax and is given next. This syntax includes some options for formatting the appearance of the chart.

```
proc gchart data = combo;
pie dmdeduc2 /value=arrow legend=discrete;
run;
```

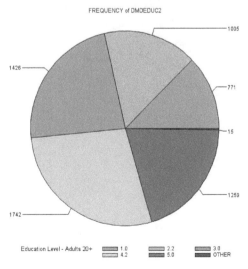

Figure 4 .9

This brief introduction to PROC SGPLOT will allow you to use charts and graphs to examine your data. The charts and graphs displayed by default are generally sufficient for internal examination of the data. However, if charts and graphs need to be used in published material or presented to persons outside the data analysis team, formatting the appearance of the chart or adding data labels, titles, or notes may be desirable.

If you are interested in learning more about how to create high-quality graphs using PROC SGPLOT I recommend the 2010 SAS Users Group International paper by S. J. Slaughter and L. D. Delwiche, "Using PROC SGPLOT for Quick High-Quality Graphs" (paper 154–2010).

Lesson #4—Study Questions

1. Which chart/graph types are best for continuous data?
2. Which chart/graph types are best for categorical data?

Lesson 5

Importing Data, Part 1

Learning Outcomes—Students Will:

- ▶ Use DATALINES to enter data into SAS manually,
- ▶ Use the Import Wizard to import Excel datasets into SAS,
- ▶ Use the Import Wizard to import Access datasets into SAS, and
- ▶ Use the Import Wizard to convert STATA and SPSS files to SAS data files.

Types of Data Files

Data rarely comes to the analyst as a SAS file. Although SAS is a very useful data management analysis tool, its data entry functions are clunky, and therefore it is rarely used for this purpose. Many researchers choose to use Excel or Access for their data entry functions and SAS for data management and to produce reliable statistics and high-quality graphs.

In addition, SAS files may be very large and difficult to transport through electronic media. For this reason, many national datasets are saved on the NCHS website as text files that must later be imported into SAS for analysis or transport files that are similar to a zipped folder in other applications. Getting data into SAS is not always intuitive. We will cover several different methods for importing data. Please note that this is not an exhaustive list but should cover the majority of file types you are likely to encounter.

Manually Entering Data into SAS—The DATALINES Option

The easiest way to get data into SAS is by entering the data manually into the program editor in a DATA step. This is a useful skill for working with small datasets or frequency tables but would be terribly inefficient for working with large databases. The following is a line listing containing data from six subjects.

Table 5.1

Subject Number	Sex	Score 1	Score 2	Project Grade
12	M	82	81	A
15	M	87	83	A
3	F	91	84	B
19	F	81	86	B
22	F	93	91	A
18	F	87	85	C

Let's look at the data more closely. We have a total of six subjects in the dataset, and we've collected five pieces of information (variables) on them: Subject Number, Sex, Score on Assignment #1, Score on Assignment #2, and Project Grade. Three of the variables are numeric (Subject Number, Score #1, and Score 2). Two of the variables are categorical (Sex and Project Grade). We will first enter the data by converting the character variables into numeric variables. For Sex, we will use 1 = Male, 2 = Female, and for Project Grade, we will use 1 = A, 2 = B, 3 = C, 4 = D, and 5 = E.

First, we create a DATA step by specifying the step command DATA. This is followed by the name you wish to give the dataset.

Second, we use an INPUT statement to specify the variables contained in the dataset we are creating. Variable names may be up to 32 characters in length and can consist of any combination of alphabetic or numeric characters. Names cannot contain any spaces or special characters, except the underscore, and must begin with a letter or underscore (no numerals). You should choose variable names that are informative but short. You may have to type the variable names many times during data manipulation and analysis, choosing names that are easy to remember and easy to type is desirable. For this exercise, I will use the following variable names:

- Subject Number = Subject
- Sex = Sex
- Score 1 = Score1
- Score 2 = Score2
- Project Grade = Project

You may use different names if you prefer.

Note 5.1 ▼

The command CARDS may also be used in place of DATALINES. This is an older keyword that harkens back to the days when data was literally typed onto 3 × 5 cards and fed into the computer to be read. Both commands are used today by programmers.

Third, we enter the command DATALINES followed by the data. A few formatting items should be noted when entering data. (1) Data should be entered one subject per line. (2) Data must be delimited. By default, SAS expects space delimitation. That is, when it reads a space, it moves onto the next variable. (3) No semicolons are used until all data has been entered.

My syntax for this dataset looks like this:

```
data testnum;
input subject sex Score1 Score2 Project;
datalines;
12 1 82 81 1
15 1 87 83 1
3 2 91 84 2
19 2 81 86 2
22 2 93 91 1
18 2 87 85 3
;
run;
```

Figure 5.1

My dataset is named TESTNUM. Because I have not specified a LIBREF, this will be saved only in the work library. If I wish to save this dataset permanently, I must specify libref.testnum.

If I highlight and run this section of code, I should see confirmation in the log that the dataset has now been created. This message confirms that I have created a dataset with six observations and five variables.

```
NOTE: The data set WORK.TESTNUM has 6 observations and 5 variables.
NOTE: DATA statement used (Total process time):
      real time           0.07 seconds
      cpu time            0.04 seconds
```

Figure 5.2

Using the explorer window, I can confirm that the dataset has been created and is in the work library.

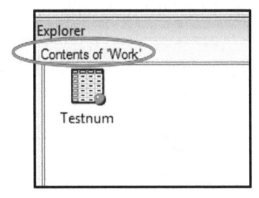

Figure 5.3

I should next print the dataset to ensure that the data was entered correctly. Because the dataset is so small, I will print the entire thing. In a larger dataset, I would use the (OBS=XX) option to look at the dataset.

```
proc print data = testnum;
run;
```

The SAS System

Obs	subject	sex	Score1	Score2	Project
1	12	1	82	81	1
2	15	1	87	83	1
3	3	2	91	84	2
4	19	2	81	86	2
5	22	2	93	91	1
6	18	2	87	85	3

Figure 5.4

I can now run PROC CONTENTS, PROC FREQ, and/or PROC MEANS on the dataset to examine the data as needed. I will run one PROC FREQ and one PROC MEANS on the data for practice. The output is included next.

```
proc freq data = testnum;
tables sex;
run;
```

The FREQ Procedure

sex	Frequency	Percent	Cumulative Frequency	Cumulative Percent
1	2	33.33	2	33.33
2	4	66.67	6	100.00

Figure 5.5

```
proc means data = testnum;
var score1;
run;
```

The MEANS Procedure

Analysis Variable : Score1				
N	Mean	Std Dev	Minimum	Maximum
6	86.8333333	4.7504386	81.0000000	93.0000000

Figure 5.6

Entering Character Data with DATALINES

Although we will primarily handle data as we did earlier, when converting character data to numeric, it's important that you have some exposure to using character variables in SAS. This time, we will enter SEX and PROJECT as characters (i.e., M/F for sex and A, B, C, D, E for grade).

The only thing different about entering character data is that I have to tell SAS which variables will be characters. Note that the default is for variables to be numeric, so SAS assumes a variable is numeric unless you specify it as a character variable. To avoid problems reading in the data, SAS will not accept character values entered into a variable it has been coded as numeric.

```
data test;
input subject sex Score1 Score2 Project;
datalines;
12 m 82 81 A
15 M 87 83 A
3 F 91 84 B
19 F 81 86 B
22 F 93 91 A
18 F 87 85 C
;
run;
```

Changing the values to M and F for sex and letter grades for project grades without specifying these as character variables will result in the following error.

Note that I have changed the data values to characters but made no other changes to the dataset.

```
24    data test;
25    input subject sex Score1 Score2 Project;
26    datalines;

NOTE: Invalid data for sex in line 27 4-4.
NOTE: Invalid data for Project in line 27 12-12.
RULE:    ----+----1----+----2----+----3----+----4----+----5----+----6----+----7----+----8----+--
27            12 m 82 81 A
subject=12 sex=. Score1=82 Score2=81 Project=. _ERROR_=1 _N_=1
NOTE: Invalid data for sex in line 28 4-4.
NOTE: Invalid data for Project in line 28 12-12.
28            15 M 87 83 A
subject=15 sex=. Score1=87 Score2=83 Project=. _ERROR_=1 _N_=2
NOTE: Invalid data for sex in line 29 3-3.
NOTE: Invalid data for Project in line 29 11-11.
29            3 F 91 84 B
subject=3 sex=. Score1=91 Score2=84 Project=. _ERROR_=1 _N_=3
NOTE: Invalid data for sex in line 30 4-4.
NOTE: Invalid data for Project in line 30 12-12.
30            19 F 81 86 B
subject=19 sex=. Score1=81 Score2=86 Project=. _ERROR_=1 _N_=4
NOTE: Invalid data for sex in line 31 4-4.
NOTE: Invalid data for Project in line 31 12-12.
31            22 F 93 91 A
subject=22 sex=. Score1=93 Score2=91 Project=. _ERROR_=1 _N_=5
NOTE: Invalid data for sex in line 32 4-4.
NOTE: Invalid data for Project in line 32 12-12.
32            18 F 87 85 C
subject=18 sex=. Score1=87 Score2=85 Project=. _ERROR_=1 _N_=6
NOTE: The data set WORK.TEST has 6 observations and 5 variables.
NOTE: DATA statement used (Total process time):
      real time           0.04 seconds
      cpu time            0.03 seconds


33    ;
34    run;
```

Figure 5.7

To correct these errors, I use the "$" (dollar sign) following the variables that should be characters. Note that in the code that follows, $ follows sex and project. Variables not followed by a $ will still be coded as numeric.

```
data test;
input subject sex Score1 Score2 Project;
datalines;
12 m 82 81 A
15 M 87 83 A
3 F 91 84 B
19 F 81 86 B
22 F 93 91 A
18 F 87 85 C
;
run;
proc print data = test;
run;
```

I can check the log to confirm that it executed without a problem. As expected, I have six observations and five variables.

```
38    data test;
39    input subject sex $ Score1 Score2 Project $;
40    datalines;

NOTE: The data set WORK.TEST has 6 observations and 5 variables.
NOTE: DATA statement used (Total process time):
      real time           0.01 seconds
      cpu time            0.00 seconds
```

Figure 5.8

Running PROC CONTENTS, I can now see the variables that are numeric or character. Recall that NUM = numeric and CHAR = character.

Alphabetic List of Variables and Attributes

#	Variable	Type	Len
5	Project	Char	8
3	Score1	Num	8
4	Score2	Num	8
2	sex	Char	8
1	subject	Num	8

Figure 5.9

Further, if I print the data, I can see that the underlying data values are retained as letters for character variables and numerals for numeric variables.

The SAS System

Obs	subject	sex	Score1	Score2	Project
1	12	m	82	81	A
2	15	M	87	83	A
3	3	F	91	84	B
4	19	F	81	86	B
5	22	F	93	91	A
6	18	F	87	85	C

Figure 5.10

A Caution for Character Variables

Now, let me illustrate one of the potential hazards of using character variables. Recall that I earlier stated that the reason we primarily rely on numeric variables is because character variables are very specific to capitalization, spacing, and spelling. In this dataset, Gender is a character variable. If you look closely at the data I entered, for five of the six subjects, I used a capital M or capital F to indicate gender; however, for the first subject, I failed to capitalize the letter M, entering it as "m." If I run PROC FREQ on this data, you can see that as far as SAS is concerned, the variable gender has three levels F, M, and m. As you might imagine, this can cause problems in data analysis, particularly in a larger dataset where such mistakes are likely to go unnoticed. For this reason, numeric variables are widely preferred for most epidemiologic applications.

The FREQ Procedure

sex	Frequency	Percent	Cumulative Frequency	Cumulative Percent
F	4	66.67	4	66.67
M	1	16.67	5	83.33
m	1	16.67	6	100.00

Figure 5.11

Using DATALINES for Fixed-Column-Width Data

In the previous example, we used spaces to delimit the data. In most data entry applications, this will work without a problem. On occasion, it is necessary to specify the width of data so that SAS can correctly read it. This is particularly true with character data, such as names and addresses. Addresses, for example, may include spaces, but still be one piece of data. For example, Salt Lake City is the name of a single city, but it has three parts and would be considered three different pieces of data if we relied on space delimitation. I will first illustrate fixed-column width using the dataset from earlier and then apply it to a more complex situation.

Fixed-column-width data is necessary when you want to, or need to, specify the maximum number of digits or characters that an individual variable may have. This is also used when reading in text and other types of data files.

Going back to the line listing from the grades data, we can see that in this dataset, the subject number can be two digits, gender one character, and Exam 1 could probably be three digits (if someone got 100%), so we'll allow for that. Exam 2 could, again, be three digits and homework grade one character.

Table 5.2

Subject Number	Gender	Exam 1	Exam 2	Homework Grade
12	M	82	81	A
15	M	87	83	A
3	F	91	84	B
19	F	81	86	B
22	F	93	91	A

To enter fixed-column-width data, two steps are required:

1. I must specify the text columns to be read for each variable. This is done in the input statement following the variable name.
2. I must enter the data so that it is correctly aligned to those columns.

Let's look at how columns are counted. Each digit, character, or space is counted separately. First, I have added in the appropriate zeros to align the data correctly. You may also use spaces to align the data. Note that here I do not need to have the spaces that originally separated the data. You may use spaces with fixed-column width but be sure to count the spaces when labeling your columns.

```
data test;
input subject sex $ Score1 Score2 Project $;
datalines;
12M8281A
15M8783A
03F9184B
19F8186B
22F9391A
18F8785C
;
run;
```

The first variable subject is found in columns 1 and 2, so I'll add that to the syntax.

```
data test;
input subject 1-2 sex $ Score1 Score2
Project $;
datalines;
12M8281A
15M8783A
03F9184B
19F8186B
22F9391A
18F8785C
;
run;
```

Next, I count sex as column 3 and then score 1 as columns 4, 5, etc. The final syntax is as follows:

```
data test;
input subject 1-2 sex $3 Score1 4-5 Score2
6-7 Project $ 8;
datalines;
12M8281A
15M8783A
03F9184B
19F8186B
22F9391A
18F8785C
;
run;
```

If I had retained the spaces, the syntax would look like this:

```
data test;
input subject 1-2 sex $4 Score1 6-7 Score2 9-10
Project $ 12;
datalines;
12 M 82 81 A
15 M 87 83 A
03 F 91 84 B
19 F 81 86 B
22 F 93 91 A
18 F 87 85 C
;
run;
```

PROC PRINT reveals that the data is entered as expected.

```
proc print data = test;
run;
```

The SAS System

Obs	subject	sex	Score1	Score2	Project
1	12	M	82	81	A
2	15	M	87	83	A
3	3	F	91	84	B
4	19	F	81	86	B
5	22	F	93	91	A
6	18	F	87	85	C

Figure 5.12

Let's look at a more complex example. Recall that the very first time we used DATALINES, we relied on spaces to delimit the data and signal to SAS that we were moving on to a new variable.

Let's consider the following line listing. For simplicity, we will deal only with two subjects.

Table 5.3

Name	Age	City	State
Oliver Schabenberger	33	Richmond	MI
John T. Smith	37	New York	NY

If the data were entered into DATALINES as it appears in the line listing, the SAS dataset would be riddled with errors. First and last names would be considered separate pieces of data, as would New and York in the city name New York. Let's see what happens.

If I run and print the dataset, I get the following output.

```
data names;
input name $ age city $ state $;
datalines;
Oliver Schabenberger 33 Richmond MI
John T. Smith         37 New York NY
;
run;
proc print data = names;
run;
```

Obs	name	age	city	state
1	Oliver	.	33	Richmond
2	John	.	Smith	37

Figure 5.13

As you can see, only the first names are recorded. Age appears as missing because the last names were read as age, and these characters were invalid values for the numeric variable age, etc.

```
'9    data names;
)0      input name $ age city $ state $;
)1      datalines;

IOTE: Invalid data for age in line 82 8-20.
)ULE:        ----+----1----+----2----+----3----+----4----+----5----+----6----+----7----+----8----+--
)2            Oliver Schabenberger 33 Richmond MI
lame=Oliver age=. city=33 state=Richmond _ERROR_=1 _N_=1
IOTE: Invalid data for age in line 83 6-7.
)3            John T. Smith         37 New York NY
lame=John age=. city=Smith state=37 _ERROR_=1 _N_=2
IOTE: The data set WORK.NAMES has 2 observations and 4 variables.
IOTE: DATA statement used (Total process time):
      real time           0.00 seconds
      cpu time            0.00 seconds
```

Figure 5.14

Here is where fixed-column width is particularly useful.

Table 5.4

Name	Age	City	State
Oliver Schabenberger	33	Richmond	MI
John T. Smith	37	New York	NY

Since Oliver Schabenberger is the longest name is the dataset, we will use this to determine how wide the columns need to be. The longest value for each variable should determine the width of the column.

```
data names;
input name $1-20 age city $25-32 state $;
datalines;
Oliver Schabenberger 33 Richmond MI
John T. Smith        37 New York NY
;
run;
proc print data = names;
run;
```

Now when we print the data, we see that the data has been correctly read.

The SAS System

Obs	name	age	city	state
1	Oliver Schabenberger	33	Richmond	MI
2	John T. Smith	37	New York	NY

Figure 5.15

Remember that the data must be correctly aligned as noted earlier. If the data is misaligned, it will be misread or deleted.

Entering Multiple Observations per Row in DATALINES

Unless otherwise specified, SAS expects one line of code after the DATALINES statement to be for a single subject. This can potentially lead to very long programs or text files. You may specify the SAS option @@ following the variable list to indicate that you are entering multiple observations per row. Here I have three variables x, y, and z, and I have a total of six observations (three per row).

```
data demo;
input x y z @@;
datalines;
1 1 12.4 1 2 11.3 1 3  1.4
2 1  2.1 2 2 19.4 1 3 10.0
;
run;
proc print data = demo;
run;
```

When reading multiple observations per row, SAS will expect x, y, z, x, y, z, x, y, z, etc., all the way to the end of the dataset. Printing the dataset shows you how the previously entered dataset looks when output.

The SAS System

Obs	x	y	z
1	1	1	12.4
2	1	2	11.3
3	1	3	1.4
4	2	1	2.1
5	2	2	19.4
6	1	3	10.0

Figure 5.16

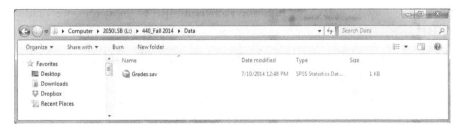

Figure 5.17

Using the Import Wizard to Convert SPSS and STATA Files into SAS

SPSS and STATA are two (of the many) other types of statistical analysis software. From time to time, it may be desirable to convert one of these types of files into a SAS data file. Luckily, SAS has made this very easy with the use of the Import Wizard. I will show this demonstration with an SPSS file, but the same procedure would apply to a STATA or JMP file.

The first step is to save your data file to your computer or the cloud where you can access it. Be sure that you know what the file is named and where it is located. Here you can see that my data file is saved on my flash drive in a data folder and called Grades.sav (.sav is the file extension for an SPSS file).

Figure 5.18

To access the Import Wizard from the main SAS toolbar, select FILE →IMPORT DATA.

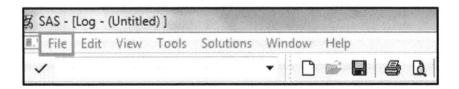

Figure 5.19

The Import Wizard will open, and you will see the following window:

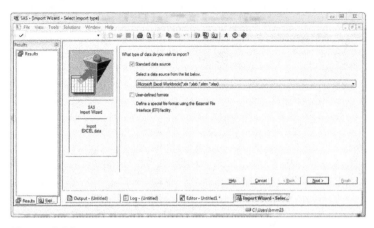

Figure 5.20

From the drop-down menu, select the file type you would like to import, in this case an SPSS file (.sav). Then select NEXT.

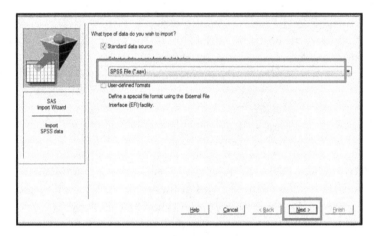

Figure 5.21

Once you push the next button, the following window will open and ask you to find where the file is saved.

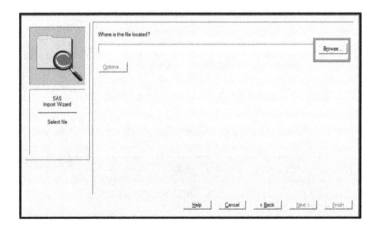

Figure 5.22

Click the browse button and navigate to your data file.

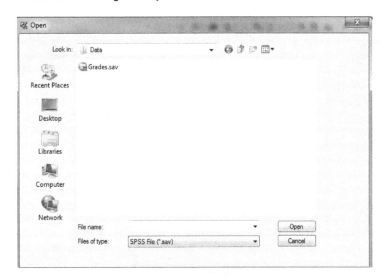

Figure 5.23

Double-click the data file, and SAS will populate the pathway. Once the pathway is populated, the "Next" button will become operational. Select NEXT.

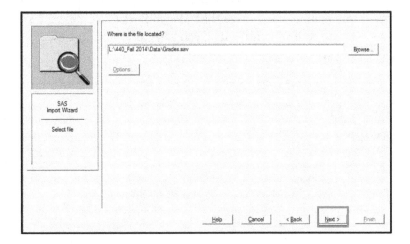

Figure 5.24

Once you select NEXT, SAS will open a window asking you where you would like to save your new SAS dataset once it's created. From the first drop-down bar, select the library you wish to save your dataset it. **Please note:** Only opened libraries will be on the drop-down menu, which means that if I want to save a dataset to a permanent library, I must open the library using a libname or the point-and-click method *before* using the Import Wizard.

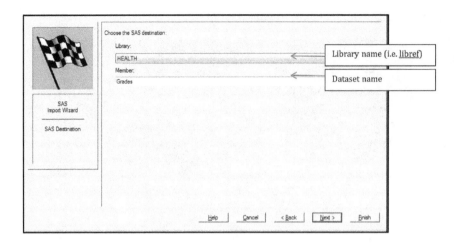

Figure 5.25

Here you can see I am saving my SAS dataset in my HEALTH library and using the dataset name GRADES (therefore the data call will be data = health.grades). After specifying the library and the dataset name, select NEXT.

The next window that comes up is an optional, but wise, choice. Behind the scenes of the Import Wizard, SAS is creating syntax that imports the data. The many steps of the Import Wizard often condense into relatively few lines of codes. This window asks you if you wish to save these PROC IMPORT statements. I always recommend that you save them. Saving the PROC IMPORT statements means that should your SAS data file be lost or corrupted, you can easily reimport the data without having to repeat the Import Wizard steps.

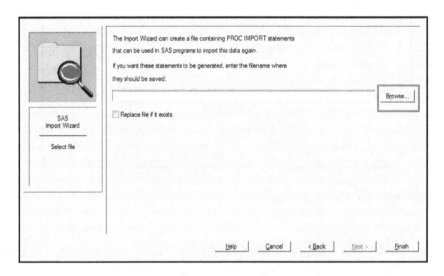

Figure 5.26

To save the import statement, click browse and navigate to where you want to save your .sas file.

Figure 5.27

In the filename blank (where it currently says *.sas), remove the asterisk (*) and write in a name for your program. I'll call mine Grades SPSS.sas. Type in the file name and slick Save.

Figure 5.28

Back in this window, click Finish.

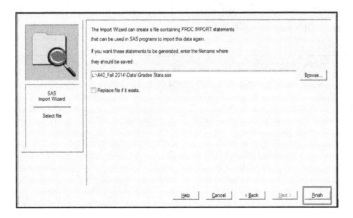

Figure 5.29

Checking Your Import

Now that the import is complete, I need to check that it was successful. Go to the SAS log, and you should see a message like this:

```
NOTE: The import data set has 6 observations and 5 variables.
NOTE: HEALTH.GRADES data set was successfully created.
```

Figure 5.30

This tells me that the dataset was successfully imported, and the number of observations and variables found in the dataset are correct. You should examine the number of observations and variables carefully; make sure that it seems logical for the dataset. If the number of variables or observations is off, this is indicative of an import that has errors.

At this point, you should also use PROC CONTENTS and perhaps other PROCs to check that the data is as you expect.

Is My Datafile Saved?

If you specified a permanent library as I did (HEALTH), then when you go to the folder on your computer or flash drive where that library is, it should allow you to see that you have a new SAS dataset (designated as .sas7bdat). Here you can see my dataset is named GRADES.sas7bdat, as I would expect since this is the libref.datasetname I used (Health.Grades).

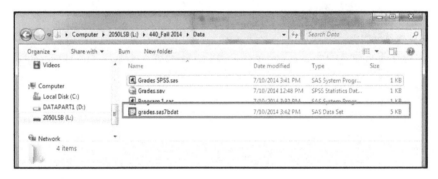

Figure 5.31

Viewing Your Saved IMPORT Statements

In the last step of the Import Wizard, we saved the import statements in a .sas program called Grades SPSS.sas. If I navigate to the folder where I saved these statements (a folder called Data on my flash drive), I can see the saved file: Grades SPSS.sas.

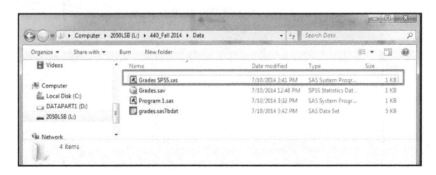

Figure 5.32

In the SAS window, select File→Open Program. Navigate to this window and double-click on Grades SPSS.sas. The following program opens:

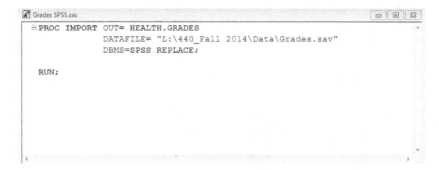

```
Grades SPSS.sas
PROC IMPORT OUT= HEALTH.GRADES
            DATAFILE= "L:\440_Fall 2014\Data\Grades.sav"
            DBMS=SPSS REPLACE;

    RUN;
```

Figure 5.33

Note that this is very few lines of code, compared to the many steps we previously executed. This is one advantage of keeping the import statements. If I need to reimport the .SAV file, I only need to highlight and run these four lines of code. This code can also be modified to change the library where the data is saved or the file name.

If you wish to import a STATA file into SAS, you would use the exact same steps, except that from the original Import Wizard window, you would need to specify a STATA file (.dta) instead of an SPSS file (.sav). All other steps remain the same.

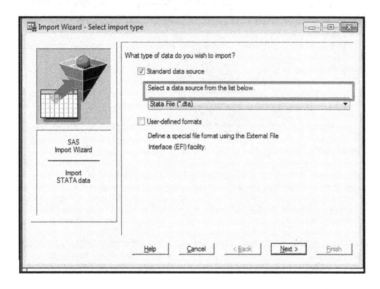

Figure 5.34

Using the Import Wizard to Import Microsoft Excel and Access Files into SAS

Converting Microsoft Excel or Access files into SAS data files is relatively easy and very similar to the previous example using SPSS. However, there is one tiny little hitch in this process that can be very confusing if you are not careful. Let's start with Excel.

Converting Excel to SAS

As we previously did, in your SAS window, go to FILE→ IMPORT DATA. The default option on the drop-down bar in this window is a Microsoft Excel Workbook. Click NEXT.

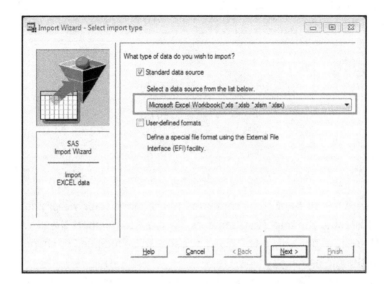

Figure 5.35

Here's where things begin to look a little different. You'll get a pop-up window like this:

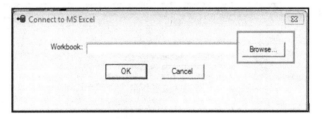

Figure 5.36

As we did before, we'll select Browse and navigate to where the .xlsx file is saved. I've saved an Excel copy of the NHANES demographics dataset in my Data folder on my flash drive as seen here:

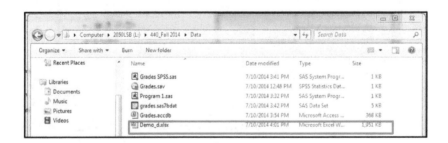

Figure 5.37

So, I'll select Browse and navigate to where my file is saved. *Notice* that when I do this, no Excel files show up in my folder! YIKES!

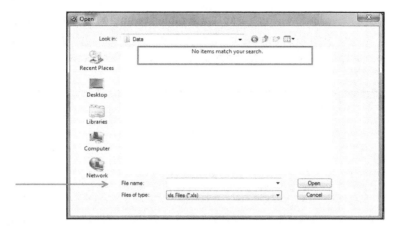

Figure 5.38

I previously showed you that my data file was nicely nestled in this folder, so what's up?!? Before you give up on the whole enterprise, there's an easy fix. Notice that in this window by default, SAS is only looking for older versions of Microsoft Excel (.xls) files. Newer versions of Excel save files as .xlsx. Change this drop-down menu to .xlsx files, and my dataset reappears. Whew!

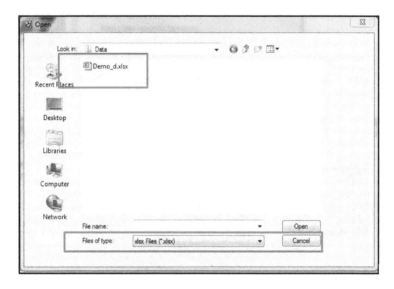

Figure 5.39

Now I can finish the import. After selecting the demo_d.xlsx file from earlier, SAS will populate the filepath:

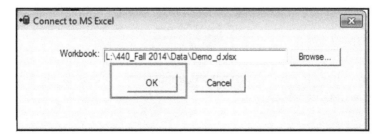

Figure 5.40

I click OK, and I get a new window, which asks which tables I want to import. Because Excel databases may have multiple "sheets" within the database, select from the drop-down menu the one you wish to import. In this case, I only have one sheet, and it is called Demo. Click NEXT, and we are in business.

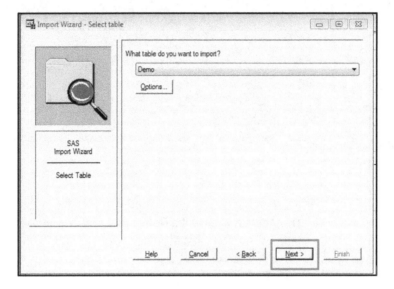

Figure 5.41

The Import Wizard will now give you this window where you can select the library and dataset name, as we previously did with the SPSS import. After entering this information, I can again save my import statements and click finish to complete the import.

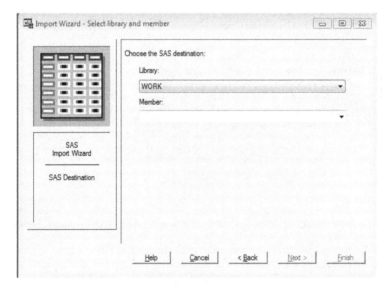

Figure 5.42

As before, I can now check my log to see that the import was successful. Here, as expected, I have 10,348 observations and 42 variables.

```
NOTE: WORK.EXCELDEMO data set was successfully created.
NOTE: The data set WORK.EXCELDEMO has 10348 observations and 42 variables.
```

Figure 5.43

Now I can view my saved import statements and use them to reimport this data file if this becomes necessary or to save a record of my work.

Converting Access Databases into SAS files

This process is very similar to the Excel files. First, I'll open the Import Wizard by going to File→Import Data and selecting Microsoft Access Database from the drop-down menu.

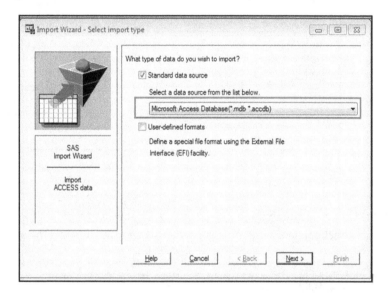

Figure 5.44

I get a slightly different window that allows me to navigate to the Access database.

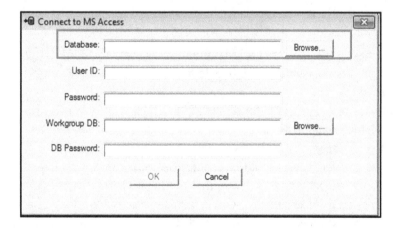

Figure 5.45

Accessing databases that have been password protected or networked will require some additional information, but we're only concerned about populating the database line. Click Browse and navigate to the database. As we saw with Excel, by default SAS, will only search for .mdb files from an old Microsoft Access file extension. Because my Access database was created in a newer version of Access, I need to change the drop-down menu to .accdb files. When I make this change, I can see my database.

Figure 5.46

After this, I select OK and navigate through the remaining windows as we did for Excel and SPSS. Once completed, I can again check to make sure my import was successful and view my import statements.

```
NOTE: WORK.GRADESACCESS data set was successfully created.
NOTE: The data set WORK.GRADESACCESS has 6 observations and 5 variables.
```

Figure 5.47

Summary

Since data rarely comes to the analyst in a SAS format, it is necessary to import other file types into SAS for analysis. In this lesson, we discussed how to import data manually from SPSS and STATA, and from Microsoft Excel and Access.

Manual data entry is simple and useful for small datasets. Data is entered using the INPUT statement in a DATA step. By default, input data is recognized as numeric. However, if entering character data, the input title for that variable must be followed by a "$" (dollar sign).

The Import Wizard is a very useful tool that makes it easy to convert databases from a variety of common programs into SAS data files. Be aware that the Import Wizard is easy but has a few quirks. Always check your imports to make sure the number of observations and variables are as you expect. It is always recommended that you save your PROC IMPORT statements generated by the Import Wizard to document your work and allow you to reimport data if necessary.

1. Why is it necessary to import data into SAS?
2. What types of data files can be imported into SAS?
3. What type of import procedure would you use when working with a small dataset?
4. Why would you import data using a fixed-column width?
5. How do I tell SAS I am entering multiple observations per row when using the DATALINES statement?
6. What does each row and column of data represent in the DATALINES statement?
7. How would you check to make sure that your data is correctly imported into SAS?
8. What is the advantage of the Import Wizard?
9. When using the Import Wizard, what information goes in the box marked "Member"?
10. When using the Import Wizard, why should you save your import program?

Lesson 6

Importing Data, Part 2

Learning Outcomes—Students Will:

> ▶ Demonstrate opening a SAS transport file in SAS using the PROC COPY,
>
> ▶ Import .txt files using SAS syntax, and
>
> ▶ Import .dat files using SAS syntax.

In the previous chapter, we manually entered data into SAS using the DATALINES command, and we used the SAS Import Wizard to convert a variety of data files into SAS data files. In this chapter, we will continue this by working with SAS transport files, ASCII (.dat) files, and text files (.txt). All of these file types will rely on manually written SAS syntax for importing into SAS.

Importing a SAS Transport File (.xpt)

A SAS transport file is a special file type designed to be easily transportable from different operating systems and different versions of SAS. A SAS transport file is compressed in much the way that a zipped file is in other applications. The smaller file size means they take up less server space when uploaded to websites and can more easily be shared from person to person. NCHS sometimes uses .xpt files, as is the case with the NHANES files we have been using up to this point.

   Importing an .xpt file will use the XPORT keyword as part of the libname statement and PROC COPY, a new procedure. We'll be working with the file hsq_d.xpt. I have this file saved in my datasets folder on my flash drive. Note that the icon for this file looks like a suitcase.

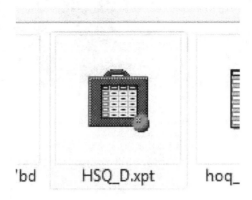

'bd HSQ_D.xpt hoq_

Figure 6.1

To import this file, I will need to write a code with TWO (2) libname statements.

- The first will point to the .xpt file and uses the XPORT command.
- The second libname statement will point to the folder where I would like to store the permanent dataset.

```
libname source XPORT 'L:\Health 440\Data\hsq_d.xpt';
libname nhanes 'L:\Health 440\Data\Data Sub Folder';
```

The first libname tells SAS where to find the .xpt file. *Notice* that in this case, the libname points to the .xpt file (hsq_d.xpt), not just to the folder containing the file. Also, note the addition of the keyword XPORT. Here I've named my library "source," but you could choose any valid library name.

```
libname source XPORT 'L:\Health 440\Data\hsq_d.xpt';
```

The second libname statement creates a library called NHANES where I can store the dataset permanently once it has been imported, and it is exactly like the libraries I have previously created. Again, you can use any valid library name.

The next step is to use the copy procedure (PROC COPY). The important keywords here are IN and OUT. IN is the input, where the data is coming from, in this case the .xpt file, and OUT is where you would like the copy to be output or saved. Think: the transport file is saved IN X, and I want you to OUTput it in Y.

The syntax then is PROC COPY IN = SOURCE, because SOURCE is the name of the library pointing to the .xpt file and OUT = NHANES, because NHANES is the name of the library where I want to save the data.

```
proc copy in=source out=nhanes;
run;
```

The full series of code looks like this:

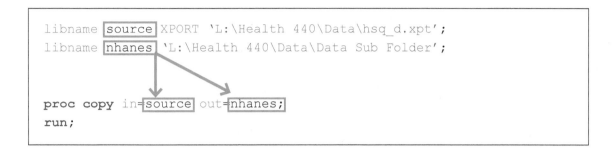

```
libname source XPORT 'L:\Health 440\Data\hsq_d.xpt';
libname nhanes 'L:\Health 440\Data\Data Sub Folder';


proc copy in=source out=nhanes;
run;
```

After executing this code, you should now see in your output library that the dataset has been converted into a standard SAS data file (.sas7bdat). The dataset can now be used as any other.

Figure 6.2

Entering Data from a Text File (.txt files)

One of the more common formats for external datasets is the text file. These are usually designated with the .txt file extension. Text files can range from very small files like this Notepad version of the previously used grades data,

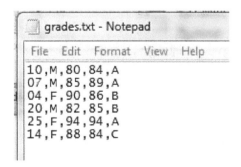

Figure 6.3

to this partial file containing data from the 2002 NSFG.

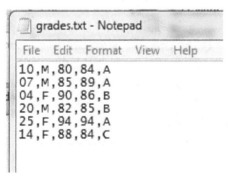

Figure 6.4

This NSFG data, which contains more than 7,000 respondents and hundreds of variables, clearly illustrates the need for an efficient means of importing this data into a SAS file.

We will import the data into SAS using two SAS statements: INPUT and INFILE. The INPUT statement works as it did with DATALINES and is followed by a list of the variables. The INFILE statement tells SAS where the .txt or .dat files are located. Before we begin, we need to know whether the data is delimited or has a fixed-column width. Delimited data uses spaces or other characters (often commas) to separate the values in the text file. Fixed-column-width data may or may not use delimiters (as illustrated in the DATALINES example). A delimiter is a partition between pieces of data such as tab, blank, symbol, or punctuation),

Let's start with the grades.txt data. An examination of this data shows that commas have been used to delimit the data. By default, SAS will assume that the data is space delimited (as we saw with our first DATALINES entry). Many text files will use comma or space delimiters; however, it is important to open your data file and check the delimiters before you proceed.

Figure 6.5

The syntax for importing a text file looks like this:

```
DATA datasetname;
INFILE 'filepath/filename';
input var1 var2 var 3;
run;
```

SAS Data Management for Public Health

- Following the DATA command, I will name my dataset gradestxt. As before, if I omit the libref, this dataset will only be saved in the work file. Unless I save a permanent copy before leaving SAS, I will have to reimport the data to use it in a new SAS session.
- INFILE is the keyword that tells SAS where the .txt file is located. The filepath is the same as we used in the libname statement, *except* that now I have to tell SAS not only which folder to look in but also which file to read. The text file is named grades.txt, and it's saved on my flash drive, which is drive L:\ on my computer. Therefore, my INFILE statement for this import will be: 'L:\Health 440\Data\grades.txt'.
- The INPUT statement is as we previously used with DATALINES and is followed by a list of the variables in the order they should be read into the data.

I just have one more step before I import the data. Recall that SAS assumes the data is space delimited unless I specify otherwise. Because my data is comma delimited, I need to tell SAS this. I do this by adding the DLM option to the INFILE statement and placing the delimiter (in this case a comma) in parentheses. Because DLM is an option, it should turn light blue, like the options we've previously seen with PROC FREQ and PROC MEANS.

```
INFILE 'filepath\filename' dlm= ',';
```

My final syntax looks like this:

```
DATA gradestxt;
INFILE 'L:\Health 440\Data\grades.txt' dlm= ',';
input subject gender $ exam1 exam2 hw_grade $;
run;
```

I can highlight and run the data, and then check the log to make sure the data imported as I expected.

```
14    DATA gradestxt;
15    INFILE 'L:\440_Fall 2012\Data Sets\grades.txt' dlm= ',';
16    input subject gender $ exam1 exam2 hw_grade $;
17    run;

NOTE: The infile 'L:\440_Fall 2012\Data Sets\grades.txt' is:
      Filename=L:\440_Fall 2012\Data Sets\grades.txt,
      RECFM=V,LRECL=256,File Size (bytes)=84,
      Last Modified=18Nov2011:14:38:34,
      Create Time=30Aug2012:15:39:32

NOTE: 6 records were read from the infile 'L:\440_Fall 2012\Data Sets\grades.txt'.
      The minimum record length was 12.
      The maximum record length was 12.
NOTE: The data set WORK.GRADESTXT has 6 observations and 5 variables.
NOTE: DATA statement used (Total process time):
      real time           0.01 seconds
      cpu time            0.00 seconds
```

Figure 6.6

From the log, I can see that everything appears to be in order. It read six records (observations) from the data, which is what I expected. Next, I need to print or otherwise check the data to make sure it looks like what I would expect. Since this dataset is small (six records), I will just print it.

```
proc print data = gradestxt;
run;
```

The data appears as I would expect, so we'll consider this a successful import.

The SAS System

Obs	subject	gender	exam1	exam2	hw_grade
1	10	M	80	84	A
2	7	M	85	89	A
3	4	F	90	86	B
4	20	M	82	85	B
5	25	F	94	94	A
6	14	F	88	84	C

Figure 6.7

One of the more common mistakes when using an INFILE statement to import data is mis-specifying the filepath or omitting the filename. When this occurs, you will get an error in SAS. A check of my log shows the following table. Remember, I'm interested primarily in the red and green text. The error "physical file does not exist" indicates that SAS could not find the file grades.txt in the specified filepath. If I correct the filepath error, I should be able to import without difficulty.

```
ERROR: Physical file does not exist, L:\440_Fall 2012\Data\grades.txt.
NOTE: The SAS System stopped processing this step because of errors.
WARNING: The data set WORK.GRADESTXT may be incomplete.  When this step was stopped there were 0
         observations and 5 variables.
WARNING: Data set WORK.GRADESTXT was not replaced because this step was stopped.
NOTE: DATA statement used (Total process time):
      real time           0.00 seconds
      cpu time            0.01 seconds
```

Figure 6.8

Importing National Survey Data from a .dat File

Now we'll work with a larger, more complex dataset. The principles illustrated in above can be used to write code to import .txt or .dat text files from any size data. Luckily, when working with National Survey Data, the .dat file is typically accompanied by a setup file. We'll work with the NSFG for this example. First, let's get the NSFG data file and the NSFG setup file from the NCHS website: http://www.cdc.gov/nchs/nsfg.htm.

On the home page, there is a menu on the left. We want the side option that says "Questionnaires, Datasets, and Related Documentation."

Figure 6.9

From this page, we'll select NSFG Cycle 6 (2002).

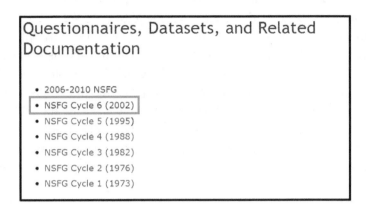

Questionnaires, Datasets, and Related
Documentation

- 2006-2010 NSFG
- NSFG Cycle 6 (2002)
- NSFG Cycle 5 (1995)
- NSFG Cycle 4 (1988)
- NSFG Cycle 3 (1982)
- NSFG Cycle 2 (1976)
- NSFG Cycle 1 (1973)

Figure 6.10

Scrolling down on the page, we'll need to download two different files. Under "Downloadable Data Files," we'll select "Female Respondent Data File."

Downloadable Data Files

- Female Respondent Data File (2002FemResp.dat)

Figure 6.11

Clicking on any file here will take you to the Data User's Agreement. You will need to accept the terms of this agreement by clicking on the "I accept these terms" link before you can continue. After you accept, you'll be directed to a file directory.

Index of /pub/Health_Statistics/NCHS/Datasets/NSFG/

Name	Size	Date Modified
[parent directory]		
.cache	0 B	9/4/98 3:37:00 PM
.cache+	0 B	9/4/98 3:37:00 PM
1973NSFGData.dat	18.9 MB	4/5/12 11:18:00 AM
1976NSFGData.dat	30.2 MB	4/5/12 11:17:00 AM
1982NSFGData.dat	31.1 MB	4/5/12 11:17:00 AM
1988FemRespData.dat	28.6 MB	4/5/12 11:17:00 AM
1988PregData.dat	53.4 MB	4/5/12 11:17:00 AM
1995FemRespData.dat	135 MB	4/5/12 11:16:00 AM
1995PregData.dat	8.5 MB	4/5/12 11:14:00 AM
2002curr_ins.dat	147 kB	11/15/10 12:05:00 PM
2002FemPreg.dat	5.8 MB	2/23/05 3:54:00 PM
2002FemResp.dat	35.9 MB	2/23/05 3:54:00 PM
2002HHvars.dat	282 kB	11/29/07 3:40:00 PM
2002Male.dat	14.0 MB	2/23/05 3:55:00 PM
2006_2010_FemPreg.dat	11.0 MB	10/12/11 10:17:00 AM
2006_2010_FemResp.dat	73.2 MB	10/12/11 10:18:00 AM
2006_2010_Male.dat	45.1 MB	10/12/11 10:18:00 AM
C5FieldworkData.asc	15.2 MB	4/6/10 7:35:00 AM
c6_curr_ins.sas7bdat	305 kB	11/15/10 12:05:00 PM
hhpartypnew.sas7bdat	305 kB	4/2/10 2:37:00 PM
HHPARTYPNEWASC.DAT	147 kB	4/2/10 2:37:00 PM
sas		4/5/12 11:20:00 AM
spss		4/5/12 11:20:00 AM
stata		4/5/12 11:21:00 AM

Figure 6.12

We will select the 2002FemResp.dat file from the list. Right-click on the file and select "Save link as." Navigate to the file on your computer or flash drive where you wish to store the data and save it. It is strongly recommended that you do NOT change the name of this file. We will also need to download the program statements. You can get these from this same file directory by opening the sas/ folder and right-clicking to save the files you need for setup. Note that the files needed depend on the year of data you are looking for. For the 2002 data, we'll need the 2002FemRespInput.sas. For most years of the data, a single "Setup" file is available. For other years and with other datasets, you may see labels and formats in separate files. For example, 2002FemRespValueLabel.sas and 2002FemRespVarLabel are formats and labels for the 2002 NSFG Female Respondent Data. These are NOT necessary for importing the data file, and if desired, they can always be added later. In a future lesson, we'll discuss how to use labels and formats. For now, let's just use the 2002FemRespInput.sas file.

Index of /pub/Health_Statistics/NCHS/Datasets/NSFG/sas/

Name	Size	Date Modified
[parent directory]		
1973FemRespSetup.sas	145 kB	4/5/12 11:20:00 AM
1976FemRespSetup.sas	568 kB	4/5/12 11:20:00 AM
1976PregSetup.sas	20.2 kB	4/5/12 11:20:00 AM
1982FemRespSetup.sas	354 kB	4/5/12 11:20:00 AM
1982PregSetup.sas	118 kB	4/5/12 11:20:00 AM
1988FemRespSetup.sas	598 kB	4/5/12 11:20:00 AM
1988PregSetup.sas	248 kB	4/5/12 11:20:00 AM
1995FemPregFile.sas	108 kB	4/5/12 11:20:00 AM
1995FemRespSetup.sas	1.4 MB	4/5/12 11:20:00 AM
2002FemPregInput.sas	6.5 kB	11/15/04 2:15:00 PM
2002FemPregValueLabel.sas	20.3 kB	11/15/04 2:22:00 PM
2002FemPregVarLabel.sas	17.3 kB	11/15/04 12:48:00 PM
2002FemRespInput.sas	78.2 kB	11/15/04 2:11:00 PM
2002FemRespValueLabel.sas	137 kB	11/15/04 4:37:00 PM
2002FemRespVarLabel.sas	220 kB	11/15/04 2:18:00 PM
2002MaleInput.sas	50.4 kB	11/15/04 2:07:00 PM
2002MaleValueLabel.sas	99.4 kB	11/15/04 2:22:00 PM
2002MaleVarLabel.sas	160 kB	11/15/04 2:18:00 PM
2006_2010_FemPregSetup.sas	66.3 kB	10/12/11 10:18:00 AM
2006_2010_FemRespSetup.sas	646 kB	10/12/11 10:18:00 AM
2006_2010_MaleSetup.sas	496 kB	10/12/11 10:18:00 AM

Figure 6.13

We'll start with the SAS program file. Open SAS and go to File → Open Program. Navigate to where you saved the program, double-click on, or highlight, the program 2002FemRespInput.sas, and click Open.

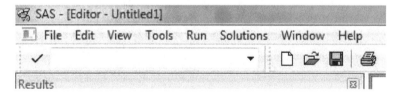

Figure 6.14

The program should open in the editor. This is by far the longest program we've worked with to date, so let's take a little time to explore it. Some of the code will be familiar to you, and other parts will be new.

At the very top of the program, you will see some commented text (denoted by the green text). These comments often give you some instructions about how to modify the program for importing the data.

Figure 6.15

```
/* This SAS program was automatically generated by Stat/Transfer */

/* The following line should contain the complete path and name of your raw data file */
FILENAME datafile 'FemResp.dat' ;

/* The following line should contain the directory in which you wish your SAS file to be stored */
libname library '.' ;

/* The following line contains the name (without the extension) for your SAS dataset */
%LET dataset = FEMALE ;
```

Figure 6.16

We will use this section to modify the code for importing. We will be setting up a couple of shortcuts here that we will explore further as the course continues.

The first part we're interested in says

```
/* The following line should contain the complete path and name of
your raw data file */
FILENAME datafile 'FemResp.dat' ;
```

We will modify this with the filepath and filename for our data. As with the transport files, this should direct all the way to the file, not just the folder. My data is saved here:

Figure 6.17

So I will modify the filepath\filename to be 'L:\NSFG 2002\2002FemResp.dat.'

```
FILENAME datafile 'L:\NSFG 2002\2002FemResp.dat.' ;
```

Next, we will modify the libname statement to where we would like to store the dataset permanently. This works just like the libname statements we have used previously. Where the original code says LIBRARY, change this to the libref you would like to use.

```
/* The following line should contain the directory in which you wish
your SAS file to be stored */
libname library '.' ;
```

Adding my library filepath modifies the code as follows:

```
libname NSFG 'L:\NSFG 2002' ;
```

Notice that I have replaced the word library with my library name (NSFG) and entered the filepath for my library between the quotation marks.

The third statement is what my final SAS dataset will be called. Here they have named the dataset FEMALE. If this is acceptable to me, I can leave it; if you want to name it something else, modify it here. Because I want to save this permanently, I need to add the libref for the library I created in the previous step.

```
/* The following line contains the name (without the extension) for
your SAS dataset */
%LET dataset = FEMALE;
Modify this statement to look like this:
%LET dataset = NSFG.FEMALE;
```

Once I have made those modifications, I am ready to run the program. However, before we do that, let's take a moment to understand what else the program contains. Scrolling down in the program, the next part we see is the DATA step. This section should look largely familiar to you. Note the use of the INFILE and INPUT statements.

```
DATA &dataset ;
 INFILE datafile LRECL=4927;
 INPUT
    CASEID $ 1-12          RSCRINF  13          RDORMRES  14
    ROSTSCRN  15           RSCREENHISP  16      RSCREENRACE  17
    AGE_A  18-19           AGE_R  20-21         CMBIRTH  22-25
    AGESCRN  26-27         MARSTAT  28          FMARSTAT  29
    FMARIT  30             EVRMARRY  31         HISP  32
    HISPGRP  33            NUMRACE  34          ROSCNT  35
    HPLOCALE  36           MANREL  37           FL_RAGE  38
    FL_RRACE  39           FL_RHISP  40         GOSCHOL  41
    VACA  42               HIGRADE  43-44       COMPGRD  45
```

Figure 6.18

The INFILE data file statement is telling SAS that the file is in the location setup earlier in the FILE-NAME shortcut. This is a shortcut for doing the same thing we did in the DATALINES examples earlier in this lesson. The option LRECL refers to the logical record length. This is automatically generated and should not be changed.

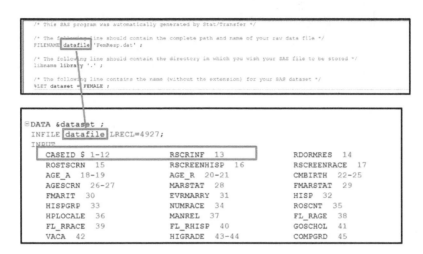

Figure 6.19

The INPUT statement is followed by the variable names in the order they appear in the dataset. Note that fixed-column width is being used. For example, the variable CASEID is a character variable found in columns 1–12; RSCRINF is a numeric variable found in column 13, etc.

Finally, at the end of the program, we will find the trusty RUN statement.

```
SECO_R   4891              SEST   4892-4893         CMINTVW   4894-4897
    CMLSTYR   4898-4901       SCREENTIME $ 4902-4909    INTVLNGTH   4910-4927 ;

RUN;
```

Figure 6.20

Since we will be running the entire program, we can just hit the running man in the toolbar to execute the whole program. I can check the log to confirm that the import was successful. I will also confirm that the dataset is now present in the NSFG library and run PROC CONTENTS to see that the dataset contains what I would expect.

The log reports that 7,643 records were read and that there were 3,087 variables. This seems reasonable, so let's proceed.

```
NOTE: The infile DATAFILE is:
      Filename=L:\NSFG 2002\2002FemResp.dat.,
      RECFM=V,LRECL=4927,File Size (bytes)=37664704,
      Last Modified=29May2013:15:40:08,
      Create Time=29May2013:15:39:36

NOTE: 7643 records were read from the infile DATAFILE.
      The minimum record length was 4927.
      The maximum record length was 4927.
NOTE: The data set NSFG.FEMALE has 7643 observations and 3087 variables.
NOTE: DATA statement used (Total process time):
      real time          32.19 seconds
      cpu time           0.90 seconds
```

Figure 6.21

Data is present in the NSFG library.

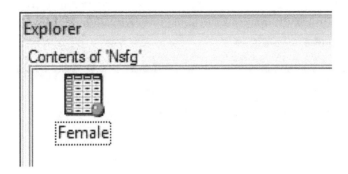

Figure 6.22

```
proc contents data=nsfg.female;
run;
```

Scrolling through PROC CONTENTS, I see that the data appears to have imported correctly.

The SAS System

The CONTENTS Procedure

Data Set Name	NSFG.FEMALE	Observations	7643
Member Type	DATA	Variables	3087
Engine	V9	Indexes	0
Created	Wednesday, May 29, 2013 04:18:38 PM	Observation Length	24704

Figure 6.23

Summary

A SAS transport files (.xpt) can be imported using two libname statements. The first statement designates the location where the data will be stored. The second statement points to the location of the .xpt file and uses the XPORT option. The PROC COPY statement copies the .xpt file and converts the data to a standard SAS data file in the location designated in the first libname statement.

Text files can be imported using the INFILE and INPUT statements in a DATA step. The INFILE statement designates the location of the file, and the INPUT statement specifies the titles of the variables. If the data is delimited with something other than spaces, it must be expressed in the INFILE statement with the DLM option.

Importing .dat files requires downloading raw datasets and their accompanying program statements. The program setup file must be modified using SAS to direct the program to the location of your downloaded .dat file.

1. Explain how a SAS transport file differs from a standard SAS data file.
2. Explain the difference between the two libname statements when importing a transport file.
3. Which procedure is used to convert a transport file into a SAS data file?
4. Explain the purpose of the INFILE statement within the DATA step.
5. Explain the use of the INPUT statement within the DATA step.

Lesson 7

Merging, Concatenation, and Managing Variables

Learning Outcomes—Students Will:

- ▶ Use a DATA step to make a copy of a SAS dataset;

- ▶ Merge two datasets, merge with tags, merge more than two datasets, and merge different types of datasets;

- ▶ Add observations to datasets; and

- ▶ Drop and keep necessary variables from datasets.

So far, we successfully imported data into SAS, and we've used PROC CONTENTS, FREQ, and MEANS to look at the data. In this lesson, we are going to start talking about how to manipulate the data, how to make copies of datasets, and how to change the structure of the datasets. In the upcoming lessons, we'll switch our focus to manipulating variables, but first, we'll deal with the datasets.

Data New; Set Old;

Repeat to yourself: Data New; Set Old;. This line of SAS code will quickly become one of your most used lines of syntax. Anytime we wish to manipulate a dataset, we want to be sure that we don't overwrite our original copy. Doing so could prove disastrous. The ugly truth is simply that at some point in the process, we will make mistakes. We'll accidentally delete observations we meant to keep, we will code over variables we needed, or we will create some other data management nightmare. For this reason, before we make any changes to a dataset, we will always start with a copy. Beginning with a copy means that when (not if) I make that disastrous mistake, I can simply revert back to my original dataset and save myself a lot of heartburn.

In Chapter 1, we were first introduced to the DATA step and its basic syntax, which looks like this:

```
data newdatasetname; set olddatasetname;
/*In this section ("within the data step"), you will type
all of your cleaning, recoding or data management
statements*/
run;
```

The DATA step is everything from DATA to RUN. Sometimes your DATA step will be small (like this one); sometimes your DATA step will be pages and pages, depending on how many data management steps you have. For this section, we will focus on only the most basic DATA step. Later in this chapter, we'll expand our DATA steps to merge and concatenate data files and manipulate our lists of variables. In subsequent chapters, we'll look at variables, but right now, all we care about is the very first line.

```
data newdatasetname; set olddatasetname;
run;
```

This line of codes says that I want to open a dataset named X (newdatasetname), and SET it equal to Y (olddatasetname). This is similar to taking a document and saving it as a new file before you edit it. First, I need to run a libname statement to open my original dataset.

```
libname nhanes 'L:\440\DataSets\Workbook';
```

I want to use the demographics dataset, so I'll confirm that the file demo_d is in my NHANES library.

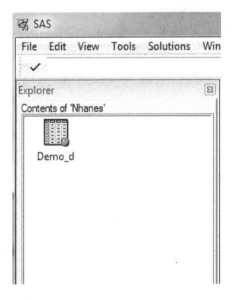

Figure 7.1

SAS Data Management for Public Health

Now, let's make a copy of it. I'm going to call my new dataset RECODE, and I would like to save it in the same library, so I'll use the compound dataset name nhanes.recode.

```
data nhanes.recode; set nhanes.demo_d;
run;
```

Now it follows the data new; set old; structure. As before, if I do not wish to save my data file permanently, I can omit the libref and specify: data recode; set nhanes.demo_d; run; which will still copy my dataset, but only save it to the work folder. This is probably sufficient for the majority of dataset copies you will make in this class.

Once I execute this command by highlighting it and clicking on the running man icon, I need to check my log to make sure the copy was successful.

```
5      data nhanes.recode;set nhanes.demo_d;
6      run;

NOTE: There were 10348 observations read from the data set NHANES.DEMO_D.
NOTE: The data set NHANES.RECODE has 10348 observations and 42 variables.
NOTE: DATA statement used (Total process time):
      real time           1.80 seconds
      cpu time            0.00 seconds
```

Figure 7.2

Note that the log shows me the code it just executed in the black text and then in the blue text, it tells me that my dataset NHANES.RECODE has 10,348 observations and 42 variables. Because I'm just making a copy, I would expect that both the old and new datasets would have the same number of observations and variables as we see here. At this point, I can also confirm that my new dataset has been saved in my NHANES library.

Figure 7.3

Again, if I had omitted the libref, this file would be visible in the WORK library. Now that I've copied my dataset, we'll learn about other file manipulations. We'll come back to Data New; Set Old; soon.

Combining Datasets

It is often necessary to combine datasets prior to analysis. There are two ways that datasets can be combined: (1) merging and (2) concatenating. Merging adds additional variables on the same subjects to the datasets, making the dataset wider. Concatenation adds additional subjects to the dataset, making the dataset longer.

Merging Data

To keep the datasets a manageable size, NHANES divides different interview modules into different physical files. The following shows just some of the more than 40 subsets of data available for each year of NHANES.

Data File Name	Doc File	Data File	Date Published
Acculturation	ACQ_D Doc	Data [XPT - 285 KB]	March, 2008
Alcohol Use (Ages 20+)	ALQ_D Doc	Data [XPT - 337 KB]	March, 2008
Allergy	AGQ_D Doc	Data [XPT - 1.80 MB]	June, 2008
Audiometry	AUQ_D Doc	Data [XPT - 1.28 MB]	April, 2009
Blood Pressure and Cholesterol	BPQ_D Doc	Data [XPT - 770 KB]	November, 2007
Bowel Health	BHQ_D Doc	Data [XPT - 262 KB]	November, 2007
Cardiovascular Disease	CDQ_D Doc	Data [XPT - 408 KB]	November, 2007
Current Health Status	HSQ_D Doc	Data [XPT - 887 KB]	January, 2008
Depression Screener	DPQ_D Doc	Data [XPT - 460 KB]	April, 2008
Dermatology	DEQ_D Doc	Data [XPT - 324	May, 2008

Figure 7.4

Each dataset within a certain year contains data on the same individuals but records different pieces of information in defined topic areas. For example, the acculturation dataset includes six questions about acculturation and respondent identification.

- SEQN - Respondent sequence number
- ACQ005 - CHECK ITEM
- ACD010A - English usually spoken at home
- ACD010B - Spanish usually spoken at home
- ACD010C - Other language(s) usually spoken at home
- ACD040 - Language(s) usually spoken at home

Figure 7.5

In practice, it is often desirable to combine data from different modules for your analysis. Let's look at an example. Suppose that I want to examine the relationship between general health status and income level. The variable to assess annual household income (INDHHINC) is in the demographics dataset (demo_d), and the variable to assess general health status is in the current health status dataset (hsq_d). To conduct this analysis, I need to get both variables in the same dataset. Merging will help me do this.

Simple Merging

1. The first step to merging is to identify the variable that the two datasets you are merging have in common. In the case of NHANES, since each row of data is about a unique individual, we are looking for the respondent ID number. In this situation, where the common variable is the respondent ID number, you will usually find this variable near the beginning of the dataset. In both the demographics (demo_d) and current health status datasets (hsq_d), the respondent identifier is the first variable SEQN. This type of merge is a 1:1 merge because each row in dataset1 matches up to one row in dataset2.

- SEQN - Respondent sequence number

Figure 7.6

2. The second step is to make sure that the merging variable has the same variable name in each dataset. In the NHANES case, both datasets use SEQN. However, elsewhere, you may encounter situations where this is not the case. For example, if you were merging on ID number and one dataset used the variable name ID and the other used IDnum, you would need to change one or both of them in the SAS DATA step before proceeding with the merge.
3. Run PROC CONTENTS on both of the datasets you would like to merge. It's important that I know the datasets well before I merge so that I can determine if the merge was completed as I expected.

```
proc contents data = nhanes.hsq_d;
proc contents data = nhanes.demo_d;
run;
```

From the PROC CONTENTS, I find that the HSQ_D dataset has n = 9,440 observations and 12 variables.

Data Set Name	NHANES.HSQ_D	Observations	9440
Member Type	DATA	Variables	12
Engine	V9	Indexes	0
Created	Thursday, October 11, 2012 05:05:07 PM	Observation Length	96
Last Modified	Thursday, October 11, 2012 05:05:07 PM	Deleted Observations	0
Protection		Compressed	NO
Data Set Type		Sorted	NO
Label	Current Health Status		
Data Representation	WINDOWS_64		
Encoding	wlatin1 Western (Windows)		

Figure 7.7

From the PROC CONTENTS for the DEMO_D dataset, I find that this dataset has n = 10,348 observations and 42 variables.

Data Set Name	NHANES.DEMO_D	Observations	10348
Member Type	DATA	Variables	42
Engine	V9	Indexes	0
Created	Tuesday, June 11, 2013 06:00:58 PM	Observation Length	336
Last Modified	Tuesday, June 11, 2013 06:00:58 PM	Deleted Observations	0
Protection		Compressed	NO
Data Set Type		Sorted	YES
Label			
Data Representation	WINDOWS_32		
Encoding	wlatin1 Western (Windows)		

Figure 7.8

In the case of NHANES, there are a total of 10,348 observations, and the modules are asked of a subset of those, either a reduced sample or a specific target population, so all of the people in the hsq_d dataset are in the demo_d dataset, but not vice versa. Therefore, I am expecting that following the merge, the dataset will have 10,348 observations and 53 variables (12 + 42 − 1 = 53). There will only be 53 (instead of 54 variables) because the common variable (SEQN) will not be duplicated.

4. Sort the data. To correctly merge, both datasets need to be sorted by the common variable. Therefore, I will first sort the datasets as follows:

```
proc sort data = nhanes.demo_d;
by seqn;
proc sort data = nhanes.hsq_d;
By seqn;
run;
```

Failure to sort the data will result in the following error and a failure of the merge.

```
ERROR: BY variables are not properly sorted on data set NHANES.DEMO_D.
SEQN=41474 HSD010=4 HSQ470=0 HSQ480=0 HSQ490=0 HSQ500=2 HSQ510=2 HSQ520=2 HSQ571=2 HSQ580=.
HSQ590=2 HSQUEX=2 SDDSRVYR=4 RIDSTATR=2 RIDEXMON=2 RIDAGEYR=16 RIDAGEMN=200 RIDAGEEX=200
RIDRETH1=4 DMQMILIT=. DMDBORN=1 DMDCITZN=1 DMDYRSUS=. DMDEDUC3=11 DMDEDUC2=. DMDSCHOL=2 DMDMARTL=5
DMDHHSIZ=5 DMDFMSIZ=3 INDHHINC=6 INDFMINC=2 INDFMPIR=0.31 RIDEXPRG=2 DMDHRGND=2 DMDHRAGE=37
DMDHRBRN=. DMDHREDU=. DMDHRMAR=3 DMDHSEDU=. SIALANG=1 SIAPROXY=2 SIAINTRP=2 FIALANG=1 FIAPROXY=2
FIAINTRP=2 MIALANG=1 MIAPROXY=2 MIAINTRP=2 AIALANG=1 WTINT2YR=6174.551667 WTMEC2YR=6487.262494
SDMVPSU=1 SDMVSTRA=47 FIRST.SEQN=1 LAST.SEQN=1 _ERROR_=1 _N_=10272
NOTE: The SAS System stopped processing this step because of errors.
NOTE: There were 9440 observations read from the data set NHANES.HSQ_D.
NOTE: There were 9395 observations read from the data set NHANES.DEMO_D.
WARNING: The data set NHANES.HSQDEMO may be incomplete.  When this step was stopped there were
        10271 observations and 53 variables.
WARNING: Data set NHANES.HSQDEMO was not replaced because this step was stopped.
NOTE: DATA statement used (Total process time):
      real time         2.10 seconds
      cpu time          0.01 seconds
```

Figure 7.9

5. Completing the merge. A DATA step is used to complete the merge using the following syntax.

```
data combineddata; merge data1 data2;
by commonvariable;
run;
```

In this case, I am still creating a new dataset, but because I am combining two different datasets instead of making a copy of one dataset, I'll use keyword MERGE. To merge hsq_d and demo_d datasets together, my specific syntax will be:

```
data nhanes.hsqdemo; merge nhanes.hsq_d nhanes.demo_d;
by seqn;
run;
```

Note that I've used a libref for my combined dataset so that a permanent copy of the combined data will be saved.

After running the code, I check the log for any errors related to the merge.

```
NOTE: There were 9440 observations read from the data set NHANES.HSQ_D.
NOTE: There were 10348 observations read from the data set NHANES.DEMO_D.
NOTE: The data set NHANES.HSQDEMO has 10348 observations and 53 variables.
NOTE: DATA statement used (Total process time):
      real time         0.66 seconds
      cpu time          0.03 seconds
```

The log tells me that SAS registered 9,440 observations from the nhanes.hsq_d dataset and 10,348 from the nhanes.demo_d dataset. Then it tells me that my new dataset NHANES.HSQDEMO has 10,348 observations and 53 variables, as expected. Running PROC CONTENTS on the combined dataset allows me to look in more detail at the combined dataset. Here I can also see that the combined dataset has been added to my library.

Figure 7.10

If I were to run PROC CONTENTS on this data, I would see that the dataset included the variables from both the demographics and the HSQ datasets.

Merging with Tags

As we saw in the earlier example, sometimes you have datasets where there is not a match for every row. There are lots of reasons why the two datasets may have a different number of observations. When this is the case, and you merge as I did in the previous step, you may end up with lots of missing data. Here I would expect that 908 persons would have missing data on the health status questions from the HSQ_D dataset (10,348 – 9,440 = 908). Using tags during merging creates a dataset in which observations are merged only under certain conditions.

Tags are behind-the-scenes flags that SAS uses to identify specific records. In this case, I am interested in tagging all of the observations in the hsq_d dataset so that my combined dataset includes only observations for which I have information.

Tags are created using the SAS keyword (IN =) followed by some word or letter. Most programmers opt for a single letter, although you may choose anything you like. I will use (IN = a).

Using the following syntax, I simultaneously tag the HSQ dataset and merge the two datasets together. Note the two changes to the code:

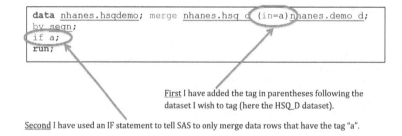

First I have added the tag in parentheses following the dataset I wish to tag (here the HSQ_D dataset).

Second I have used an IF statement to tell SAS to only merge data rows that have the tag "a".

Figure 7.11

After running the merge code, I check my log and now see that the combined dataset has only 9,440 observations, but still 53 variables.

```
NOTE: There were 9440 observations read from the data set NHANES.HSQ_D.
NOTE: There were 10348 observations read from the data set NHANES.DEMO_D.
NOTE: The data set NHANES.HSQDEMO has 9440 observations and 53 variables.
NOTE: DATA statement used (Total process time):
      real time            0.65 seconds
      cpu time             0.00 seconds
```

Figure 7.12

The use of tags can be expanded to include additional criteria. For example, if I want to combine datasets where both have observations that are not in the other, I could use the following code:

```
data nhanes.hsqdemo; merge nhanes.hsq_d (in=a)
nhanes.demo_d (in=b);
by seqn;
if a and b;
run;
```

Here I am tagging datasets HSQ_D with an "a" and DEMO_D with a "b." The IF statement now tells SAS only to retain records that are in BOTH a AND b. Similarly, you may use the OR keyword to have SAS keep records that are in a OR b. For example,

Keep all records in dataset2 and only those that match in dataset1.

```
data combined; merge dataset1 dataset2 (in=b);
if b;
```

Keep only records that exist in both datasets.

```
data combined; merge dataset1 (in=a)dataset2 (in=b);
if a and b;
```

Keep only records that are in dataset1 and not in dataset2.

```
data combined; merge dataset1 (in=a)dataset2 (in=b);
if a and not b;
```

Merging More than Two Datasets

Merging three or more datasets follows the same procedure as a simple merge. As long as each dataset contains the common variable, you can merge them in a single DATA step. For example, here I am merging four datasets together:

```
data nhanes.hsqdemo; merge nhanes.hsq_d nhanes.hoq_d
nhanes.ffqraw_d nhanes.demo_d;
by seqn;
run;
```

Merging 1:M to data

In some cases, you may wish to perform a more complicated merge where you combine two different types of data. For example, suppose that I have individual-level data that I am interested in analyzing, but I want to include some variables about the places people live from the US Census. I can use an identifier such as STATE FIPS Code to merge data about Utah with each of the people in the dataset who live in Utah. The process for merging is very similar, except that in this case, I do not have a unique set of data for each individual, as there may be many people within a dataset who share a state of residence. Here is a hypothetical example:

Suppose I have the following datasets:

The SAS System

Obs	FIPS	sex	health
1	Utah	Male	Good
2	Utah	Male	Fair
3	Utah	Female	Poor
4	Idaho	Male	Fair
5	Idaho	Female	Good
6	Idaho	Female	Poor
7	Wyoming	Female	Fair
8	Wyoming	Male	Good
9	Wyoming	Male	Poor
10	Utah	Female	Fair

Figure 7.13 The first dataset contains three variables:

1. State FIPS code a unique identifier used to designate individual US states
2. Sex = Respondent's Gender
3. Health = Respondent's Health Status as Good, Fair, or Poor

**Please note that although you are seeing the formatted versions of the datasets (we'll discuss formats soon!), the variables are all numeric. The datasets without the formats look like this:

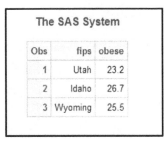

Obs	STATE	sex	health
1	49	1	1
2	49	1	2
3	49	2	3
4	16	1	2
5	16	2	1
6	16	2	3
7	56	2	2
8	56	1	1
9	56	1	3
10	49	2	2

Obs	fips	obese
1	49	23.2
2	16	26.7
3	56	25.5
4	32	22.6

The second data set has 2 variables:
1) State FIPS code.
2) Obese = the percentage of persons in each state who are obese.

The SAS System

Obs	fips	obese
1	Utah	23.2
2	Idaho	26.7
3	Wyoming	25.5

Figure 7.14

In this case, I'm interested in knowing if there is a relationship between an individual's perception of health (Health Status) and the level of obesity in his or her state of residence. To merge these two datasets together, complete the following steps:

1. Find the common variable. Here the common variable is FIPS.
2. Make sure the common variable is the same in both datasets. Here the variable is denoted FIPS in both cases.
3. Run PROC CONTENTS.
4. Sort the data by the common variable.

```
proc sort data = person;
by fips;
run;
proc sort data = state;
by fips;
run;
```

After the sort, we can see that both datasets are ordered Idaho, Utah, and then Wyoming.

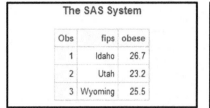

The SAS System

Obs	fips	obese
1	Idaho	26.7
2	Utah	23.2
3	Wyoming	25.5

The SAS System

Obs	FIPS	sex	health
1	Idaho	Male	Fair
2	Idaho	Female	Good
3	Idaho	Female	Poor
4	Utah	Male	Good
5	Utah	Male	Fair
6	Utah	Female	Poor
7	Utah	Female	Fair
8	Wyoming	Female	Fair
9	Wyoming	Male	Good
10	Wyoming	Male	Poor

Figure 7.15

5. Complete the merge.

```
data health; merge person state;
by fips;
run;
```

For demonstration purposes, we will now print the merged data file to show the results.

The SAS System

Obs	FIPS	sex	health	obese
1	Idaho	Male	Fair	26.7
2	Idaho	Female	Good	26.7
3	Idaho	Female	Poor	26.7
4	Utah	Male	Good	23.2
5	Utah	Male	Fair	23.2
6	Utah	Female	Poor	23.2
7	Utah	Female	Fair	23.2
8	Wyoming	Female	Fair	25.5
9	Wyoming	Male	Good	25.5
10	Wyoming	Male	Poor	25.5

Figure 7.16

Please note that all of the persons from Idaho have the same value for obese (here the statewide prevalence of obesity).

Let's use a similar example to review tagging. The first dataset is identical; however, now the second dataset contains information for a fourth state: Nevada. Let's see what happens when we merge without tagging.

Obs	FIPS	sex	health
1	Utah	Male	Good
2	Utah	Male	Fair
3	Utah	Female	Poor
4	Idaho	Male	Fair
5	Idaho	Female	Good
6	Idaho	Female	Poor
7	Wyoming	Female	Fair
8	Wyoming	Male	Good
9	Wyoming	Male	Poor
10	Utah	Female	Fair

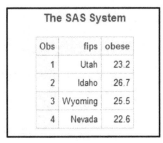

The SAS System

Obs	fips	obese
1	Utah	23.2
2	Idaho	26.7
3	Wyoming	25.5
4	Nevada	22.6

Figure 7.17

```
proc sort data = person;
by fips;
run;
proc sort data = state;
by fips;
run;


data health; merge person state;
by fips;
run;


proc print data = health;
run;
```

We see in the merged dataset that there is missing data next to Nevada because there were no respondents to the health questionnaire who lived in Nevada.

Obs	FIPS	sex	health	obese
1	Idaho	Male	Fair	26.7
2	Idaho	Female	Good	26.7
3	Idaho	Female	Poor	26.7
4	Nevada	.	.	22.6
5	Utah	Male	Good	23.2
6	Utah	Male	Fair	23.2
7	Utah	Female	Poor	23.2
8	Utah	Female	Fair	23.2
9	Wyoming	Female	Fair	25.5
10	Wyoming	Male	Good	25.5
11	Wyoming	Male	Poor	25.5

Figure 7.18

By adding the tagging criteria, we eliminate this problem.

```
data health; merge person (in=a) state;
by fips;
if a;
run;

proc print data = health;
run;
```

Obs	FIPS	sex	health	obese
1	Idaho	Male	Fair	26.7
2	Idaho	Female	Good	26.7
3	Idaho	Female	Poor	26.7
4	Utah	Male	Good	23.2
5	Utah	Male	Fair	23.2
6	Utah	Female	Poor	23.2
7	Utah	Female	Fair	23.2
8	Wyoming	Female	Fair	25.5
9	Wyoming	Male	Good	25.5
10	Wyoming	Male	Poor	25.5

Figure 7.19

Merging with Rename

If your common variables do not have the same variable name, you can rename them as part of the merge step. Using the state and health data from earlier, let's assume that in the first dataset with the respondent information, the state variable is called PLACE, and in the second dataset with the state-level data, the state variable is called FIPS.

Obs	STATE	sex	health
1	Utah	Male	Good
2	Utah	Male	Fair
3	Utah	Female	Poor
4	Idaho	Male	Fair
5	Idaho	Female	Good
6	Idaho	Female	Poor
7	Wyoming	Female	Fair
8	Wyoming	Male	Good
9	Wyoming	Male	Poor
10	Utah	Female	Fair

Figure 7.20

Since I need these to be the same to allow the merge, I must rename them. I could do this in a separate DATA step or as part of the merge. This syntax will help.

```
data combined; merge dataset1
(rename=(oldvar=commonvar)) dataset2;
by commonvar;
run;
```

Adjusting this to be specific to my case I get the following:

```
data health; merge person(rename=(place=fips)) state;
by fips;
run;
```

Obs	STATE	sex	health
1	Utah	Male	Good
2	Utah	Male	Fair
3	Utah	Female	Poor
4	Idaho	Male	Fair
5	Idaho	Female	Good
6	Idaho	Female	Poor
7	Wyoming	Female	Fair
8	Wyoming	Male	Good
9	Wyoming	Male	Poor
10	Utah	Female	Fair

Obs	fips	obese
1	Utah	23.2
2	Idaho	26.7
3	Wyoming	25.5
4	Nevada	22.6

Figure 7.21

The SAS System

Obs	fips	sex	health	obese
1	Idaho	Male	Fair	26.7
2	Idaho	Female	Good	26.7
3	Idaho	Female	Poor	26.7
4	Nevada	.	.	22.6
5	Utah	Male	Good	23.2
6	Utah	Male	Fair	23.2
7	Utah	Female	Poor	23.2
8	Utah	Female	Fair	23.2
9	Wyoming	Female	Fair	25.5
10	Wyoming	Male	Good	25.5
11	Wyoming	Male	Poor	25.5

Figure 7.22

The data were successfully merged, and the common variable name is now FIPS.

Concatenation

Concatenation is the process of adding observations to a dataset. Let's stick with the same datasets we've been using, but let's assume that we now have an additional dataset from the eastern states region in addition to our Utah, Idaho, and Wyoming data.

We now have the original dataset (left) and one containing data from Virginia, West Virginia, and Maryland (right).

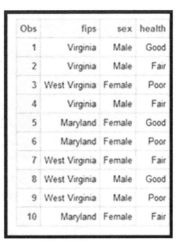

Figure 7.23

To concatenate, at least some of the variables must be the same. Each dataset contains the same three variables: FIPS, SEX, and HEALTH. No sorting or common identifier is needed. The following syntax concatenates the dataset:

```
data new; set dataset1 dataset2;
run;
```

Specific to my situation, I will use the following syntax:

```
data health2; set person east;
run;
```

We note from the log that there were ten observations read from each of the datasets, and the new dataset now has 20 observations and 3 variables:

```
NOTE: There were 10 observations read from the data set WORK.MANY.
NOTE: There were 10 observations read from the data set WORK.EAST.
NOTE: The data set WORK.HEALTH2 has 20 observations and 3 variables.
NOTE: DATA statement used (Total process time):
      real time            0.01 seconds
      cpu time             0.01 seconds
```

Figure 7.24

Printing the dataset shows us that the additional observations have been appended to the bottom of the first dataset.

Obs	fips	sex	health
1	Utah	Male	Good
2	Utah	Male	Fair
3	Utah	Female	Poor
4	Idaho	Male	Fair
5	Idaho	Female	Good
6	Idaho	Female	Poor
7	Wyoming	Female	Fair
8	Wyoming	Male	Good
9	Wyoming	Male	Poor
10	Utah	Female	Fair
11	Virginia	Male	Good
12	Virginia	Male	Fair
13	West Virginia	Female	Poor
14	Virginia	Male	Fair
15	Maryland	Female	Good
16	Maryland	Female	Poor
17	West Virginia	Female	Fair
18	West Virginia	Male	Good
19	Wyoming	Male	Poor
20	Maryland	Female	Fair

Figure 7.25

If one dataset contains additional variables, those variables will be missing for any observations where data was not available. In this example, the dataset containing the eastern states' information had another variable called OTHER. When we concatenate, respondents from Utah, Idaho, and Wyoming have missing data for the OTHER variable.

Obs	fips	sex	health	other
1	Utah	Male	Good	.
2	Utah	Male	Fair	.
3	Utah	Female	Poor	.
4	Idaho	Male	Fair	.
5	Idaho	Female	Good	.
6	Idaho	Female	Poor	.
7	Wyoming	Female	Fair	.
8	Wyoming	Male	Good	.
9	Wyoming	Male	Poor	.
10	Utah	Female	Fair	.
11	Virginia	Male	Good	1
12	Virginia	Male	Fair	1
13	West Virginia	Female	Poor	1
14	Virginia	Male	Fair	1
15	Maryland	Female	Good	1
16	Maryland	Female	Poor	1
17	West Virginia	Female	Fair	1
18	West Virginia	Male	Good	1
19	Wyoming	Male	Poor	1
20	Maryland	Female	Fair	1

Figure 7.26

Dropping or Keeping Variables

DROP and KEEP are keywords that can be used to restrict the variables retained in a dataset. datasets with a large number of variables may take a long time to process commands. DROP and KEEP allow you to exclude (DROP) or retain (KEEP) only the variables you need in the dataset and thereby improve processing times. Other reasons for reducing the number of variables in a dataset may include removing personal identifiers from a dataset being used for research purposes. DROPping the identifying variables provides a de-identified dataset, which helps protect the privacy of survey respondents. You may use either DROP or KEEP, but not both in the same DATA step.

DROP and KEEP are used in a DATA step and can be used independently (i.e., in their own DATA step) or as part of a DATA step merge or concatenation (or any other data management task). The generic syntax looks like this:

Dropping from a single dataset:

```
data new; set old;
drop list of variables to drop;
run;
```

Note that if I had other data management tasks in the DATA step, they could appear before or after the DROP statement.

Keeping from a single dataset:

```
data new; set old;
keep list of variables to keep;
run;
```

The choice of DROP versus KEEP is usually a matter of choosing whichever has the shortest list of variables. One very helpful shortcut in SAS is that contiguous variables can be denoted with two hyphens (- -). Contiguous variables are those that are next to each other in position order in the dataset.

For example, from the NHANES demographic dataset, these 15 variables are the first 15 from left to right. If I wished to DROP or KEEP the variables SEQN through race (RIDRETH1), I could use the following code:

```
data new; set old;
keep SEQN—RIDRETH1;
run;
```

Variables in Creation Order			
#	Variable	Type	Len	Label
1	SEQN	Num	8	Respondent sequence number
2	SDDSRVYR	Num	8	Data Release Number
3	RIDSTATR	Num	8	Interview/Examination Status
4	RIAGENDR	Num	8	Gender - Adjudicated
5	RIDAGEYR	Num	8	Age at Screening Adjudicated - Recode
6	RIDAGEMN	Num	8	Age in Months - Recode
7	RIDAGEEX	Num	8	Exam Age in Months - Recode
8	RIDRETH1	Num	8	Race/Ethnicity - Recode
9	DMQMILIT	Num	8	Veteran/Military Status
10	DMDBORN	Num	8	Country of Birth - Recode
11	DMDCITZN	Num	8	Citizenship Status
12	DMDYRSUS	Num	8	Length of time in US
13	DMDEDUC3	Num	8	Education level - Children/Youth 6-19
14	DMDEDUC2	Num	8	Education level - Adults 20+
15	DMDSCHOL	Num	8	Now attending school?

Figure 7.27

Let's look at a few more examples. Continuing with the NHANES demographics dataset, we will first DROP two variables, DMQMILIT and DMDBORN, using the following syntax:

To illustrate the changes in the dataset, I will run PROC CONTENTS on both the old and new datasets. You will recall that the original NHANES demographics dataset has 10,348 observations and 42 variables.

Data Set Name	NHANES.DEMO_D	Observations	10348
Member Type	DATA	Variables	42

Figure 7.28

After completing the DROP, my dataset REDUCE1 has 10,348 observations and 40 variables. This is expected since I have dropped two variables.

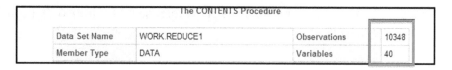

The CONTENTS Procedure			
Data Set Name	WORK.REDUCE1	Observations	10348
Member Type	DATA	Variables	40

Figure 7.29

Next, we will KEEP the variables DMQMILIT, DMDBORN, and SEQN using the following syntax:

```
Data reduced; set nhanes.demo_d;
keep dmqmilit dmdborn seqn;
run;
proc contents data = reduced;
run;
```

Again, I will run PROC CONTENTS to check the datasets. As expected, the REDUCED dataset contains 10,348 observations, but only three variables.

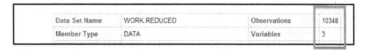

| Data Set Name | WORK.REDUCED | Observations | 10348 |
| Member Type | DATA | Variables | 3 |

Figure 7.30

The procedure for DROP or KEEP works the same regardless of the number of variables you use.

DROP and KEEP from a Merge Step

It is also possible to DROP or KEEP variables within the same DATA step you are using to merge. When merging datasets, you may only need a few of the variables from one dataset. Using DROP or KEEP allows you to merge in only what you need. The process is the same, and PROC CONTENTS can still be used to check that the DROP and KEEP steps happened as expected.

Keeping from a merge step:

```
data new;
merge dataset1 dataset2 (keep=list of variables to keep);
by commonvar;
run;
```

Dropping from a merge step:

```
data new; merge dataset1
dataset2 (drop=list of variables to drop);
by commonvar;
run;
```

Making a Data Subset

While DROP and KEEP are tools for reducing the number of variables retained in a dataset, making a subset allows you to reduce the number of observations retained in a dataset. Sometimes it is desirable to analyze just a portion of a population, such as men or adolescents, or those who are not married. Creating a subset allows you to do this. Creating the subset is done in a DATA step using an IF clause:

```
data subset; set olddataset;
if some condition;
run;
```

We can substitute any number of conditions in the IF clause using AND/OR qualifiers.

Suppose, for example, that we are interested in analyzing only those under age 55 for our research question. We can use the following code to reduce the number of observations to only include those less than or equal to age 55.

```
data subset; set nhanes.demo_d;
if ridageyr le 55;
run;
```

*where le = Less than or equal to.

We see from PROC CONTENTS that the dataset now contains a reduced number of observations (8,571) and 42 variables.

Data Set Name	WORK.SUBSET	Observations	8571
Member Type	DATA	Variables	42

Figure 7.31

If we wanted to examine only men under age 55, we could do the following:

```
data subset; set nhanes.demo_d;
if ridageyr le 55 and riagendr = 1;
run;
```

As we can see from the PROC CONTENTS, again, this has further reduced the number of observations in the dataset.

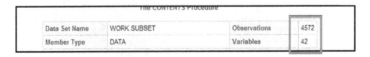

	The CONTENTS Procedure		
Data Set Name	WORK.SUBSET	Observations	4572
Member Type	DATA	Variables	42

Figure 7.32

SAS Data Management for Public Health

Summary

The DATA step is the foundation of data management in SAS. To avoid irreparable errors, we always work with new copies of datasets. Using the Data new; set old; structure for the dataset allows us to manipulate a copy of our original data, leaving the original data untouched. Typically, additional data management tasks will follow the data new; set old;.

There are several ways to combine datasets prior to analysis. Merging adds additional variables on the same subjects to the datasets, making the dataset wider. Concatenation adds additional subjects to the dataset making the dataset longer. Merging datasets requires identifying and sorting by a common variable and then merging the sets in a DATA step. Tags can be used to ensure that the merged dataset only contains observations you stipulate. Sometimes it may be necessary to rename certain variables to facilitate a merge.

Concatenation requires that at least some of the variables from each dataset be the same. A simple DATA step concatenates the datasets. If one dataset contains additional variables, those variables will be missing for any observations where data was not available.

Dropping or keeping variables is another way to manage data in your datasets. These options can be part of a DATA step with or without merging or concatenation.

Making a data subset allows the analyst to restrict the data to certain criteria, such as age, gender, marriage status, etc. A subset is created by using AND/OR qualifiers in an IF statement in the DATA step.

Lesson #7—Study Questions

1. What type of SAS step is used to manipulate datasets?
2. How do you make a copy of a dataset?
3. Why is copying a dataset a wise practice?
4. What method of combining datasets is used to add additional variables on the same subjects?
5. What method of combining datasets is used to add additional subjects to a dataset?
6. When merging two datasets, how would you go about finding the common variable?
7. When merging a dataset with 30 variables to a dataset with nine variables, why would the merged dataset only contain 38 variables?
8. Why would you use tags when merging datasets?
9. When concatenating datasets, is it necessary to sort by a common variable? Is it necessary to have common variables in the datasets?
10. Explain the difference between DROP and KEEP statements, and describe why you may use them.
11. What kind of statement would you use in the DATA step to limit your dataset to subjects under the age of 65?

Lesson 8

Data Cleaning Rules

Learning Outcomes—Students Will:

▶ Identify the principles of data cleaning in SAS and demonstrate their use,

▶ Demonstrate an understanding of the basics of IF-THEN logic code as the basis of variable manipulation in SAS, and

▶ Demonstrate using IF-THEN logic to remove invalid responses from both categorical and continuous data.

Data Cleaning

Up to this point, we have defined what "dirty data" are and how to recognize "dirty data" using a combination of data dictionaries and output from the contents, frequency, means, and univariate procedures in SAS. Once "dirty data" have been identified, we must decide what to do with them. Typically, invalid values, such as don't know and refused, will be treated as missing, as there is no meaningful respondent information included there, and I cannot guess what the value might be for a respondent. Values that appear to be out of range, illegal, or incompatible should be confirmed against the original survey or data source and corrected if possible. If correction is not possible, these values may also be treated as missing.

The Rules of Data Cleaning

The process of removing "dirty data" is called data cleaning. Ensuring that data are of good quality and as accurate as they can be is the first step of any analysis. There are five rules of data cleaning that can guide our efforts:

1. Always make new datasets.
2. Always make new variables.

3. Never delete records.
4. Always account for missing values.
5. Always check your work.

These rules are all based on the general idea that you never want to compromise your original dataset. Mistakes will be made, and it's vital that you have an original source of the data to go back to and start again if need be.

Always Make New DataSets

As discussed in the previous chapter, we can use a DATA step to create copies of our original dataset, which we can later manipulate. Creating a "new" dataset is the process of copying the current spreadsheet of data into a new spreadsheet. This is essentially the same as saving two copies of the same file, with different names on your computer. As a reminder, in its simplest form, a DATA step looks like this:

```
data newdatasetname; set olddatasetname;
/*Recoding Steps*/
run;
```

Recall that we also used a DATA step for merging and concatenation, reducing the list of variables with DROP/KEEP statements, and making data subsets to reduce the number of observations in a dataset.

The dataset name, as discussed in Lesson 1, is important in this DATA step. If you want your new dataset to be saved in your permanent library, you must give your new dataset a two-part name (libref. datasetname); otherwise, it will be saved in your temporary (work) library. This step is typically carried out in concert with the recoding of variables discussed in the next step. You may also come across circumstances where you will need to make a copy of your new dataset. To do so, you will follow the same process, essentially making a copy of the copy. The syntax would look like this:

```
data one; set original;
/*Recoding Steps*/
run;

data two; set one;
/*Recoding Steps*/
run;
```

This is extremely important! <u>Using a second DATA step in the same program that is also set to the original dataset will remove all of the changes made in your first dataset</u>. To avoid this problem, in most situations, you can complete *all* of your recoding in a single DATA step.

Always Make New Variables

As with datasets, the rationale for creating a new variable or copy is that only manipulating the copy prevents you from permanently damaging the original data. If a mistake is made (and they often are), you can always revert to the original variable and try again. In addition, using a copy of the variable allows you to check your work by comparing the old and new variables using a cross tabulation, thus ensuring accuracy in recoding. Creating new variables is done *within* the DATA step, typically the same DATA step you used to create a copy of the dataset. The simplest syntax, shown next, creates a copy of only the variable:

```
data newdatasetname; set olddatasetname;
newvariable=oldvariable;
run;
```

The step is useful *only* when you are wishing to change the name of a variable to something more user-friendly like RIDAGEYR into AGE. If you need to clean out invalid values (don't know/refused) or plan to collapse categories, those steps will automatically create copies of the variables. We'll cover this later in this lesson.

It is also possible to create a new variable that is not a copy of any previous variable in the dataset. The code is very similar:

```
data newdatasetname; set olddatasetname;
newvariable=15;
run;
```

Here I am creating a new variable that is equal to 15. This would create a new column in the dataset and have every cell be equal to 15. We could do this same process to create a new variable that was missing for all persons or was equal to any single value. Sometimes this is useful when combining datasets, as some variables may not be specified. For example, I may have a dataset that is restricted to men and a separate dataset that is restricted to women. Because the datasets only contain a single gender, there is no variable specifying gender in the datasets. If I merged these together, I may no longer be able to identify which respondents are male and which are female. Prior to merging, I may specify Gender = 1 in the Male dataset and Gender = 2 in the Female dataset. Then when I merge the datasets, the gender will still be identifiable.

Never Delete Records

Records or observations (the rows of the database) should never be deleted. Deleting records is a permanent choice that in the case of weighted data can result in problems. Unwanted values (such as "don't know" or "refused") should be set to missing. Values classified as missing in SAS will not be used in the analysis, but the integrity of the data will remain. In addition, analysis of subpopulations can be done easily in SAS, thus making the deletion of records risky and unnecessary. Note that creating subsets (as we did in Chapter 6) is not recommended when the data need to be weighted, as is true for most survey data.

Always Account for Missing Values

Depending on the software package that the data was originally constructed in, and the decisions made by those doing the data entry, missing or invalid data can be denoted in a variety of ways. It's important that you assess how the missing and invalid values are denoted prior to commencing data cleaning. The system-missing value in SAS is the period (.), but missing values may also be denoted by some other character or number. In addition to simply identifying how a missing or invalid value is denoted, you should try to understand why a value is missing. Data may be missing for a variety of reasons. Ideally, you can identify whether they are missing because someone was never asked a specific question (skip pattern) or because it was refused or overlooked. While you may be unable to account for every missing or invalid value, you should try to carefully understand the source of the numbers, particularly for variables with large numbers of missing values.

All of the missing data we've looked at in SAS up to this point has been denoted with the system-missing value of the period (.). SAS, however, also allows special types of missing, which can be useful for indicating the source or reason behind missing data. These special types of missing are created by using letters (A–Z) or an underscore (_) preceded by a period to indicate different types of missing. For example, .A to represent "unknown," .B for "refused," and .L for "Legitimate Skip." Although these designations will be visible in the codebook and when using the /MISSING and /MISSPRINT options in PROC FREQ, they will be treated as standard missings and omitted from the analysis. If special types of missing have been used, this should be noted in the codebook. Here is an example from the National Survey of Children's Health.

In this excerpt from the codebook, you can see that they are using two "special types of missing" denoted as L and M in the codebook.

How would you describe the condition of [S.C.]'s teeth?

K2Q01_D	Frequency	Percent	Cumulative Frequency	Cumulative Percent
L - LEGITIMATE SKIP	5122	5.35	5122	5.35
M - MISSING IN ERROR	10	0.01	5132	5.36
1 - EXCELLENT	46181	48.27	51313	53.63
2 - VERY GOOD	23549	24.61	74862	78.24
3 - GOOD	15895	16.61	90757	94.86
4 - FAIR	3926	4.10	94683	98.96
5 - POOR	932	0.97	95615	99.94
6 - HAS NO NATURAL TEETH	22	0.02	95637	99.96
96 - DON'T KNOW	35	0.04	95672	99.99
97 - REFUSED	5	0.01	95677	100.00

Figure 8.1

If I run a PROC FREQ on this variable using the/MISSPRINT option, I can see that these show up in my output table as missing. Note that if I need to use these in my code (i.e., I want to change to the system missing or some other value), I need to reference them as .L or .M. The preceding period indicates that these are missing values.

		How would you describe the condition of [S.C.]'s teeth?		
K2Q01_D	Frequency	Percent	Cumulative Frequency	Cumulative Percent
L	5122	.	.	.
M	10	.	.	.
1	46181	51.00	46181	51.00
2	23549	26.01	69730	77.01
3	15895	17.55	85625	94.57
4	3926	4.34	89551	98.90
5	932	1.03	90483	99.93
6	22	0.02	90505	99.96
96	35	0.04	90540	99.99
97	5	0.01	90545	100.00

Frequency Missing = 5132

Figure 8.2

Always Check Your Work

Everyone makes mistakes. It doesn't matter how long you've been coding; from time-to-time, you will make mistakes. A mistake in the coding of your data could be disastrous and lead to incorrect conclusions, if not detected. For these reasons, you must always check your work. You will check recoding work through cross tabulations of the new and old variables, always using the /MISSING or /MISSPRINT option as follows:

```
proc freq data=libref.datasetname;
tables oldvariable*newvariable /missing;
run;
```

Note 8.1 ▼

It is important to put the variables in this order, causing the old variable to be in the rows and the new variable to be in the columns of your output. As recoding often results in fewer values for the new variable, doing so provides greater ease in reading your output.

This cross tabulation will help you evaluate if everything was coded correctly. You should specifically look to ensure the following:

- All of the categories are mutually exclusive.
- All of the values are accounted for.

- Everything is coded as you would expect.
- No records were lost.

Mistakes in recoding are common, but easily fixed, so don't fret, just go back to your code, correct what went wrong, and try again.

IF-THEN Logic

Recoding is the process of cleaning out the unwanted or invalid responses from a variable and collapsing categories or otherwise manipulating variables. This is done through a series of IF-THEN statements. An IF-THEN statement goes like this:

> IF some condition is met THEN take this action.

Both the IF and THEN must be present in the statement to form a complete argument. When using IF-THEN logic, the use of ELSE and ELSE IF are also important operators. ELSE means "otherwise," such as the following:

> IF some condition is met THEN take this action;
> ELSE IF some condition is met THEN take this action.

The second statement is executed only if the first condition is not met. This improves the overall efficiency of SAS coding, as it reduces the number of observations that SAS has to read to determine if the IF condition is met.

ELSE without the IF (as in ELSE IF) says that if none of the prior conditions are met, then do this, such as the following:

> IF some condition is met THEN take this action;
> ELSE IF some condition is met THEN take this action;
> ELSE take this action.

Since IF-THEN logic is the basis for almost all recoding and data management in SAS, it's important to understand the underlying concepts beneath the code.

Mathematical Operators

SAS uses a variety of mathematical operators to complete the arguments in the if-then-else if-then statements. The following is a table of those operators and their equivalents in SAS (you can use either the symbol or text according to your preference).

Table 8.1

Expression	Symbol	TEXT	Text
Equal	=	EQ	eq
Less than	<	LT	lt
Less than or equal to	<=	LE	le
Greater than	>	GT	gt
Greater than or equal to	>=	GE	ge
Not (or negation)	^	NOT	not

Not equal to	^=	NE	ne
In this set	in( )	IN( )	in( )
Contains	?	CONTAINS	contains

These symbols and operators can be powerful tools for creating variables that result from adding or multiplying (or other mathematical functions) two or more variables. I generally prefer the text operators, simply because they require fewer keystrokes than the symbol versions. Here are some logic code examples using these mathematical operators.

> IF state of residence = Utah THEN Time Zone = Mountain;
> ELSE IF state of residence = Missouri THEN Time Zone = Central;
>
> IF age lt 18 THEN group = child;
> ELSE IF age ge 18 THEN group = adult;

The remaining operators work similarly.

Using IF-THEN Logic to "Clean Out" Invalid Responses

As we discussed in the earlier sections, data cleaning takes place in a DATA step using IF-THEN logic. Our first step is to identify which values are "invalid" for each variable we plan to clean. Let's look at some examples from the NHANES Sexual Behavior data (SXQ_D). Sexual behavior questions, by nature, are very personal and often have higher levels of question refusal than less personal questions.

This variable SXQ021 asks the 3,285 respondents to the sexual behavior questionnaire if they have ever had vaginal, anal, or oral sex. Let's look carefully at the data.

SXQ021 - Ever had vaginal, anal, or oral sex

Variable Name:	SXQ021
SAS Label:	Ever had vaginal, anal, or oral sex
English Text:	The next set of questions is about your sexual behavior. By sex, we mean vaginal, anal, or oral sex. Please remember that your answers are strictly confidential. Have you ever had vaginal, anal, or oral sex?
English Instructions:	VERBAL INSTRUCTIONS TO SP: Please select yes, no.
Target:	Both males and females 20 YEARS - 59 YEARS

Code or Value	Value Description	Count	Cumulative	Skip to Item
1	Yes	2614	2614	
2	No	309	2923	SXQ280
7	Refused	4	2927	SXQ280
9	Don't know	4	2931	SXQ280
.	Missing	354	3285	

Figure 8.3

First, we noticed that YES is coded as 1 and NO is coded as 2. We would consider either of these to be valid responses to the question, "Have you ever had sex?" Four people refused to answer this question and are coded as 7, and an additional four people said they did not know if they have ever had sex and are coded as 9. We will consider both 7 and 9 to be invalid data values. Before we proceed with our analysis, we'll want to clean these out by setting them to missing. We should also note that there are 354 missing. As always, it's important to understand why there is a large number of missing. Here we note that the target is males and females 20–59 years.

A review of the other documentation in the codebook shows 354 people did not complete the questionnaire for sexual behavior. This matches the number of missing for our variable.

SXAISC - Interview Status Code

Variable Name:	SXAISC
SAS Label:	Interview Status Code
English Text:	Interview Status Code
Target:	Both males and females 20 YEARS - 59 YEARS

Code or Value	Value Description	Count	Cumulative	Skip to Item
1	Complete	2928	2928	
2	Partial	3	2931	
3	Not done	354	3285	End of Section
.	Missing	0	3285	

Figure 8.4

Now that we understand the data, we'll use SAS to set the invalid responses to missing so that we can calculate correct proportions.

```
1 data recodes; set nhanes.sxq_d;
2 if sxq021 = 7 then hadsex = .;
3 else if sxq021 = 9 then hadsex = .;
4 else hadsex = sxq021;
  run;
```

1. Our first line of code creates a new dataset called recodes saved in the work library (you may save this permanently if you desire)—that is, a copy of the SXQ_D dataset saved in my NHANES library.
2. The second line of code sets the refused to missing *and* creates a new variable called HADSEX. Note that the first mention of a new variable always generates a new column in your dataset. For this reason, when cleaning data, it is not necessary to separately copy your variable.
3. The third line says if they didn't refuse, but they did say they didn't know, then make HADSEX equal to missing.
4. And, finally, the fourth line says if they didn't refuse and didn't say they didn't know, then make my new variable equal to the old variable. This will copy all my 1, 2, and `.`'s from the original variable to my new variable. Note the order of the variables in this step: new = old. If you inadvertently reverse this order and list old = new, all of your data (or at least a lot of it) will be missing!

Now I need to check my work to make sure that the variable is coded as I would like. We do this using a cross tabulation. The order of the variables is not strictly important, but I typically recommend old*new. The reasons for this will make more sense when you begin collapsing categories. Because you typically move from a lot of categories to fewer categories, using the old*new crosstab can make reading output easier.

```
proc freq data = recodes;
tables sxq021*hadsex/missing;
run;
```

Notice that I am using the /missing option. This is always recommended when checking recodes. I am also running this PROC FREQ on my recodes dataset. This is the dataset that has the new variable HADSEX in it. If I were to run this using the original sxq_d dataset, I would get the following error:

```
proc freq data = nhanes.sxq_d;
tables sxq021*hadsex/missing;
run;


12    proc freq data = nhanes.sxq_d;
13    tables sxq021*hadsex/missing;
ERROR: Variable HADSEX not found.    ←
14    run;
```

This usually indicates either that I have used the wrong dataset or that the variable is spelled wrong.

Figure 8.5

Running the cross tabulation on the correct dataset yields the following table:

Frequency Percent Row Pct Col Pct	Table of SXQ021 by hadsex				
	SXQ021(Ever had vaginal, anal, or oral sex)	hadsex			
		.	1	2	Total
.		354	0	0	354
		10.78	0.00	0.00	10.78
		100.00	0.00	0.00	
		97.79	0.00	0.00	
1		0	2614	0	2614
		0.00	79.57	0.00	79.57
		0.00	100.00	0.00	
		0.00	100.00	0.00	
2		0	0	309	309
		0.00	0.00	9.41	9.41
		0.00	0.00	100.00	
		0.00	0.00	100.00	
7		4	0	0	4
		0.12	0.00	0.00	0.12
		100.00	0.00	0.00	
		1.10	0.00	0.00	
9		4	0	0	4
		0.12	0.00	0.00	0.12
		100.00	0.00	0.00	
		1.10	0.00	0.00	
Total		362	2614	309	3285
		11.02	79.57	9.41	100.00

Figure 8.6

When looking at this table, I have the old variable SXQ021 in the rows and the new variable HADSEX in the columns. The old variable has five levels, including missing, and the new variable only has three levels, including missing. Next, I want to check the mapping, in this case to make sure that the 7 and 9's from the old variable are coded as missing in the new variable. Here we see that this is true and that all other values have been directly copied. What was missing originally is still missing; what was 1 before is still 1; what was 2 before is still 2.

Now that we've checked our work we can go on to our next variable. This variable SXQ101 is asking for the number of male sexual partners with whom the respondent had vaginal, anal, or oral sex.

SXQ101 - #male sex partners/lifetime

Variable Name:	SXQ101
SAS Label:	#male sex partners/lifetime
English Text:	In your lifetime, with how many males have you had vaginal, anal, or oral sex?
English Instructions:	VERBAL INSTRUCTIONS TO SP: Please enter a number or enter zero for none. ENTER NUMBER.
Target:	Females only 20 YEARS - 59 YEARS

Code or Value	Value Description	Count	Cumulative	Skip to Item
0 to 200	Range of Values	1457	1457	
77777	Refused	10	1467	
99999	Don't know	3	1470	
.	Missing	1815	3285	

Figure 8.7

The codebook indicates that this is a discrete variable with values ranging from 0 to 200 and with refused coded as 77777 and don't know as 99999. We also see that there are 1,815 people missing data on this. Because this is the first time we've encountered a discrete variable, let's look at a PROC FREQ. Typically, we use PROC MEANS for discrete or continuous variables, but showing you a PROC FREQ in this case will help you understand how the structure of the data correlates with what is shown in the codebook.

```
proc freq data = nhanes.sxq_d;
tables sxq101;
run;
```

This produces a very long table, but as you can see, there are values such as 0, 1, 2, 3, 4, 5, and 6, all the way up to 200. In this case, not all numbers between 0 and 200 appear in the table since respondents did not mention some values.

Table 8.2   #Male Sex Partners/Lifetime

SXQ101	Frequency	Percent	Cumulative Frequency	Cumulative Percent
0	11	0.75	11	0.75
1	278	18.91	289	19.66
2	146	9.93	435	29.59
3	170	11.56	605	41.16
4	113	7.69	718	48.84
5	136	9.25	854	58.10
6	100	6.80	954	64.90
7	51	3.47	1005	68.37
8	50	3.40	1055	71.77
9	28	1.90	1083	73.67
10	103	7.01	1186	80.68
11	12	0.82	1198	81.50
12	37	2.52	1235	84.01
13	6	0.41	1241	84.42
14	4	0.27	1245	84.69
15	54	3.67	1299	88.37
16	8	0.54	1307	88.91
17	1	0.07	1308	88.98
18	1	0.07	1309	89.05
19	2	0.14	1311	89.18
20	45	3.06	1356	92.24
21	3	0.20	1359	92.45
22	3	0.20	1362	92.65
23	1	0.07	1363	92.72
24	2	0.14	1365	92.86
25	29	1.97	1394	94.83
26	1	0.07	1395	94.90
27	2	0.14	1397	95.03
28	3	0.20	1400	95.24
30	17	1.16	1417	96.39
35	5	0.34	1422	96.73
36	1	0.07	1423	96.80
38	1	0.07	1424	96.87
39	1	0.07	1425	96.94
40	5	0.34	1430	97.28
42	1	0.07	1431	97.35
45	3	0.20	1434	97.55
50	10	0.68	1444	98.23
53	1	0.07	1445	98.30
60	5	0.34	1450	98.64

(Continued)

Table 8.2 #Male Sex Partners/Lifetime

SXQ101	Frequency	Percent	Cumulative Frequency	Cumulative Percent
65	1	0.07	1451	98.71
70	1	0.07	1452	98.78
90	2	0.14	1454	98.91
100	2	0.14	1456	99.05
200	1	0.07	1457	99.12
77777	10	0.68	1467	99.80
99999	3	0.20	1470	100.00

Back to the codebook:

SXQ101 - #male sex partners/lifetime

Variable Name:	SXQ101
SAS Label:	#male sex partners/lifetime
English Text:	In your lifetime, with how many males have you had vaginal, anal, or oral sex?
English Instructions:	VERBAL INSTRUCTIONS TO SP: Please enter a number or enter zero for none. ENTER NUMBER.
Target:	Females only 20 YEARS - 59 YEARS

Code or Value	Value Description	Count	Cumulative	Skip to Item
0 to 200	Range of Values	1457	1457	
77777	Refused	10	1467	
99999	Don't know	3	1470	
.	Missing	1815	3285	

Figure 8.8

Refused and don't know will be treated the same as before. We need to account for the missing: 1,815 missing observations is a large number (more than half of our total sample). Looking at the target, we notice that this question about male sex partners was only asked of female respondents. From here, the process is the same. I will continue my recoding in the same DATA step I previously used.

```
data recodes; set nhanes.sxq_d;
if sxq021 = 7 then hadsex = .;
else if sxq021 = 9 then hadsex = .;
else hadsex = sxq021;

if sxq101 = 77777 then malepartner = .;
else if sxq101 = 99999 then malepartner = .;     ←——— New Code
else malepartner = sxq101;

run;
```

Since both our old and new variables are discrete, we will use PROC MEANS or PROC UNIVARIATE to check our work. You may use PROC FREQ, but as you saw earlier, running PROC FREQ on a discrete or continuous variable produces a very large, unwieldy table. The following code will run PROC MEANS for both the old and new variables.

```
proc means data = recodes;
var sxq101 malepartner;
run;
```

This produces the following results:

Variable	Label	N	Mean	Std Dev	Minimum	Maximum
SXQ101	#male sex partners/lifetime	1470	740.8646259	7813.58	0	99999.00
malepartner		1457	7.7584077	11.2294672	0	200.0000000

Figure 8.9

Notice that the mean number of sexual partners in the original variable is 741 compared to a much more likely 7.8 in the new variable. You can also see that the maximum value in the new variable is within the valid values (200 as denoted in the codebook). We also notice that the N, which is based on nonmissing observations, changes by 13, which we would expect since we had 10 women who refused to answer this question and 3 who reported that they did not know the answer.

Summary

Cleaning data is done through the use of IF-THEN logic code within a DATA step in SAS. Cleaning data involves setting invalid responses such as don't know and refused so that the proportions and means calculated from this data are not compromised. It is important to follow good rules for data cleaning. The five rules for data cleaning include the following:

1. Always make new datasets.
2. Always make new variables.
3. Never delete records.
4. Always account for missing values.
5. Always check your work.

It is extremely important to always check your work. Good SAS coding provides opportunities for checking your work that increase programmer confidence in the results. Categorical variables are checked using cross tabulations in PROC FREQ and continuous or semicontinuous variables are checked using PROC MEANS or PROC UNIVARIATE.

1. Explain what data cleaning is?
2. Why are invalid values often set to missing?
3. What are the five rules of data cleaning?
4. Explain how you "make a new dataset."
5. Explain how you "make a new variable" in the simplest situation.
6. Explain why deleting records is risky.
7. Explain what it means to "account for missing values."
8. Describe one method of checking your work when cleaning categorical variables.
9. What type of logic code is the basis for all recoding and data management tasks?
10. Explain the difference between IF-THEN, ELSE IF-THEN, and ELSE statements.
11. Explain how you might check your work with continuous variables.
12. Fill in the expression for each of the following mathematical operators:

Table 8.3

Operator	Expression
EQ	
GT	
LE	
IN()	
NOT	
CONTAINS	
LT	

Lesson 9

Recoding Categorical Variables

Learning Outcomes—Students Will:

- ▶ Demonstrate the use of IF-THEN logic to recode one categorical variable into a new categorical variable with reduced categories, and

- ▶ Use PROC FREQ to check their work.

Simple Recoding and Data Cleaning for Categorical Variables

In the previous lesson, we used IF-THEN statements in the DATA step to clean out invalid values prior to data analysis. In this lesson, we will take that a step further by learning to collapse categories for categorical variables, called recoding. Categorical data are made up of nominal and ordinal data. Categorical data are common in public health. Many of the factors we are interested in are categorical: alive or dead, pregnant or not pregnant, male or female, white, black, Asian, Hispanic, American Indian, etc. First, we must decide how we would like our variable to look.

From the NHANES Demographics dataset, let's examine the variable DMDHSEDU—"What is the highest grade or level of school the household reference person's spouse has completed or the highest degree received?"

DMDHSEDU	Target			
	B(0 Yrs. to 150 Yrs.)			
Hard Edits	**SAS Label**			
	HH Ref Person's Spouse Education Level			

English Text: What is the highest grade or level of school {you have/NON-SP SPOUSE has completed or the highest degree {you have/he/she has} received

English Instructions:

Code or Value	Description	Count	Cumulative	Skip to Item
1	Less Than 9th Grade	793	793	
2	9-11th Grade (Includes 12th grade with no diploma)	723	1516	
3	High School Grad/GED or equivalent	1243	2759	
4	Some College or AA degree	1522	4281	
5	College Graduate or above	1249	5530	
7	Refused	4	5534	
9	Don't know	20	5554	
.	Missing	4794	10348	

Figure 9.1

From this section of the data dictionary, I can see that there are five valid categories, numbered one to five. Refused is denoted as 7 (n = 4) and don't know is denoted as 9 (n = 20). I also see that I have 4,794 missing values. This represents 46% of the total respondents (n = 10,348). The high number of missing is concerning, but from the data dictionary, I can determine that these missing values are so-called valid missing. Persons who are not married would not have been asked this question, thus a valid and reasonable skip pattern resulted in the missing data. Now that I've identified how the variable is presently coded, identified how the missing and invalid responses are denoted, and accounted for my missing, I can proceed with recoding. The first step is to determine the final categories I want to arrive at. In this case, I am interested in creating a variable that dichotomizes education into 1 = No College Education and 0 = At Least Some College.

It is often valuable to create a chart with your old and new variables to make sure you account for all of the variables in your categories.

Table 9.1

Value of New Variable	Value(s) of Old Variables
1= No college education	
0 = At least some college	
.= missing/invalid	

Now that I've created the chart (typically I would just sketch this on scratch paper), I'll go back to the codebook to decide what belongs where. Value 1 = Less than 9th grade; this clearly belongs in the No College Education group, so I'll add it there. Value 2 = 9th–11th grade; this also belongs in the No College Education group.

DMDHSEDU	Target			
	B(0 Yrs. to 150 Yrs.)			
Hard Edits	SAS Label			
	HH Ref Person's Spouse Education Level			

English Text: What is the highest grade or level of school {you have/NON-SP SPOUSE has completed or the highest degree {you have/he/she has} received

English Instructions:

Code or Value	Description	Count	Cumulative	Skip to Item
1	Less Than 9th Grade	793	793	
2	9-11th Grade (Includes 12th grade with no diploma)	723	1516	
3	High School Grad/GED or equivalent	1243	2759	
4	Some College or AA degree	1522	4281	
5	College Graduate or above	1249	5530	
7	Refused	4	5534	
9	Don't know	20	5554	
.	Missing	4794	10348	

Figure 9.2

I'll continue this process until each of the values of the original variables 1–9 are assigned to one of the three levels of my new variables: No College Education, At Least Some College, or Missing.

Table 9.2

Value of New Variable	Value(s) of Old Variables
1= No college education	1, 2, 3
0 = At least some college	4, 5
.= missing/invalid	7, 9, .

Now that I have determined how I want my new variable to look and what values of the old variable belong in each category of the new variable, I can proceed with my recoding. As we did in Lesson 7, we'll use IF-THEN logic code within a DATA step to conduct the recoding.

```
data recode; set nhanes.demo_d;
if dmdsedu=1 then nocollege1=1;
else if dmdhsedu=2 then nocollege1=1;
else if dmdhsedu=3 then nocollege1=1;
else if dmdhsedu=4 then nocollege1=0;
else if dmdhsedu=5 then nocollege1=0;
else if dmdhsedu=7 then nocollege1=.;
else if dmdhsedu=9 then nocollege1=.;
else if dmdhsedu=. then nocollege1=.;
/*Additional recodes may be included here*/
run;
```

Here I am creating a new dataset called recode (saved in my work library) from the original dataset demo_d in my permanent NHANES library. My new variable is called nocollege1. This is the simplest form of recoding where each value of the old variable is managed in a single SAS statement.

Now that I've recoded this variable, I need to check my work. I'll do this using the frequency procedure and running a cross tabulation of the oldvar*newvar with the /MISSING option.

For ease of display, I will suppress the row, column, and total percentages, although this is not strictly necessary, as follows:

```
proc freq data=recode;
tables dmdhsedu*nocollege1/missing norow nocol nopercent;
run;
```

Which results in the following:

Frequency	Table of DMDHSEDU by nocollege1				
	DMDHSEDU(HH Ref Person's Spouse Education Level)	nocollege1			
		.	0	1	Total
	.	4794	0	0	4794
	1	0	0	793	793
	2	0	0	723	723
	3	0	0	1243	1243
	4	0	1522	0	1522
	5	0	1249	0	1249
	7	4	0	0	4
	9	20	0	0	20
Total		4818	2771	2759	10348

Figure 9.3   In examining the cross tabulation, I'm looking to make sure that I still have 10,348 rows of data and that each of the categories is coded as I expect. Following the arrows in the image, I see that DMDHSEDU values 1, 2, and 3 are all coded as nocollege1 = 1. Using a similar pattern, I can see that DMDHSEDU values 4 and 5 are coded as nocollege1 = 0 and that DMDHSEDU values 7, 9, and (.) are all coded as nocollege1 =., which is what I wanted. This variable is now ready for use in the analysis.

Dealing with Multiple Recodes

As with cleaning, I can use a single DATA step to recode multiple variables. This will always be more efficient than using a new DATA step for each variable. The same is true when checking my recodes using PROC FREQ. I can add multiple cross tabulations simply by adding a space.

Summary

Recoding allows you to organize data from the codebook's method of recording to however you wish to view the data. To recode, you must first identify how the variable is presently coded, identify how the missing and invalid responses are denoted, and account for missing data before determining what final categories you wish to arrive at. Creating a chart with your old and new variables can help you make sure that you have correctly accounted for all of the variables in your categories. Don't forget to check your work with a FREQ procedure.

Lesson #9—Study Questions

1. Explain what categorical data are.
2. Explain how "recoding" differs from the simple cleaning of data.
3. Explain how you check your work.
4. Explain basic use of the IF-THEN logic code.
5. Identify the SAS step in which all recoding takes place.

Lesson 10

More Recoding Techniques

Learning Outcomes—Students Will:

▶ Demonstrate the use of the OR, IN, BETWEEN, and ELIMINATION techniques for recoding.

Alternate Recoding Techniques

In Lesson 9, we learned the simplest form of recoding categorical variables. Although this is not always the easiest or quickest way of recoding, it will always work and is the basis for recode programming. In this lesson, we'll discuss some additional recoding techniques. Each of the five recoding techniques is useful in some situations. You will quickly find your own coding preference and stick to that technique for the majority of your recodes; however, it's important that you are familiar with all of the techniques, as there will be some recoding situations in which only one of the recoding techniques will work.

Recoding Using "OR"

Recoding using OR allows you to reduce the number of SAS statements necessary for recoding by combining statements using OR. We'll use the same example from Lesson 9, where we are recoding DMDHSEDU (Spouse's Education) into 1 = No College and 0 = At Least Some College.

Table 10.1

Value of New Variable	Value(s) of Old Variables
1 = No college education	1, 2, 3
0 = At least some college	4, 5
.= missing/invalid	7, 9, .

For example, in the scenario, DMDHSEDU values 1, 2, and 3 are all being combined into nocollege2 value 1. Therefore, I can code them in a single statement, as shown here:

```
data recode; set nhanes.demo_d;
if dmdhsedu=1 OR dmdhsedu=2 OR dmdhsedu=3 then nocollege2=1;
else if dmdhsedu=4 OR dmdhsedy=5 then nocollege2=0;
else if dmdhsedu=7 OR dmdhsedy=9 OR dmdhsedy=. then nocollege2=.;
/*Additional recodes may be included here*/
run;
```

Here I am saying that if DMDHSEDU= 1, OR = 2, OR = 3 then code nocollege2 as 1, etc.

Comparing this to my code from Lesson 9, you can see that I've reduced my code from eight lines to three. This recoding technique is particularly useful when you are doing conditional recoding (where values of a new variable are dependent on the values of two or more other variables).

```
data recode; set nhanes.demo_d;
if dmdsedu=1 then nocollege1=1;
else if dmdhsedu=2 then nocollege1=1;
else if dmdhsedu=3 then nocollege1=1;
else if dmdhsedu=4 then nocollege1=0;
else if dmdhsedu=5 then nocollege1=0;
else if dmdhsedu=7 then nocollege1=.;
else if dmdhsedu=9 then nocollege1=.;
else if dmdhsedu=. then nocollege1=.;
/*Additional recodes may be included here*/
run;
```

Recoding Using "IN ( )"

Recoding using IN ( ) is perhaps the most popular and easiest of the recoding techniques. Here the mathematical operator is the word IN followed by parentheses:

```
data recode; set nhanes.demo_d;
if dmdhsedu IN (1,2,3) then nocollege3=1;
else if dmdhsedu IN (4,5) then nocollege3=0;
else if dmdhsedu IN (7,9,.) then nocollege3=.;


/*Additional recodes may be included here*/
run;
```

Here I am saying, if DMDHSEDU has a value that is either 1, 2, or 3, then code nocollege3 as 1, etc. As before, you can see that this is much quicker than simple recoding and only three lines of code.

Recoding Using BETWEEN

This recoding technique is perhaps the most difficult to master, but it is extremely useful for categorical variables with a large number of values or for continuous variables you wish to convert into categories.

Recoding the same variable as before, the example that follows uses BETWEEN to recode the variable:

```
data recode; set nhanes.demo_d;
if 1 le dmdhsedu le 3 then nocollege4=1;
else if 4 le dmdhsedu le 5 then nocollege4=0;
else if 7 le dmdhsedu le 9 then nocollege4=.;

/*Additional recodes may be included here*/
run;
```

I am using the operator "le," which means less than or equal to. The mathematical function looks like this:

$$X \leqslant \text{value of the variable} \leqslant Y.$$

Looking at the first nine observations from the data obtained using PROC PRINT, we can more closely examine this argument.

Obs	DMDHSEDU
1	2
2	.
3	4
4	.
5	4
6	5
7	4
8	3
9	5

Figure 10.1

For this first observation, which has a value of 2 for DMDHSEDU, SAS asks the question, is $1 \leqslant 2$, and is $2 \leqslant 3$? If both of these conditions are met, then the new variable, nocollege4, is coded as 1.

Similarly, for observation 6, which has a value of 5, SAS asks the question, is $1 \leqslant 5$, and is $5 \leqslant 3$? In this case, only the first condition is met, so SAS moves to the next line of code asking, is $4 \leqslant 5$, and is $5 \leqslant 5$? As this condition is met, SAS codes the new variable, nocollege4, as 0.

A similar process would continue until SAS determines which conditions each observation meets.

Recoding Using Elimination

Recoding using elimination is accounting for all but one condition and setting everything equal to some value, as follows:

```
if dmdhsedu=4 then nocollege5=0;
else if dmdhsedu=5 then nocollege5=0;
else if dmdhsedu=7 then nocollege5=.;
else if dmdhsedy=9 then nocollege5=.;
else if dmdhsedu=. then nocollege5=.;
else nocollege5=1;
```

Using if-then-else if statements, I am accounting for all of the conditions that should be coded as 0 or missing (.) in the new variable. Then in the sixth line of code, I'm saying otherwise (else) if DMDHSEDU is not 4, 5, 7, or 9, then make it 1. This kind of coding was previously shown in Lesson 8 and can be particularly helpful when you only need to change a few conditions.

Earlier, we used this example:

```
data recodes; set nhanes.sxq_d;
if sxq021 = 7 then hadsex = .;
else if sxq021 = 9 then hadsex = .;
else hadsex = sxq021;
run;
```

This is another example of recoding by elimination, where I'm asking SAS to remove all of the 7s and 9s and make them missing (.). I want to keep everything else the same as it was originally coded, so I write else newvar = oldvar (hadsex = sxq021). Just as in the prior example, if sxq021 is not 7 or 9, then it is set equal to whatever value was recorded in the original variable.

Recoding Using Multiple Techniques

SAS reads each and every statement (from keyword to semicolon) separately. There is no reason that a single recoding technique must be used for an entire variable recode. In fact, it's very typical to combine techniques. As you will quickly learn, you will have personal preferences for recoding, and there will be times where one technique is clearly superior to another. Although each will give you the same answer, in different situations, one technique may be much easier and more efficient. Feel free to mix it up as you see fit, just be sure to check your recodes to ensure that you get what you are expecting.

This example uses BETWEEN, OR, and IN to recode DMDHSEDU into nocollege6.

```
if 1 le dmdhsedu le 3 then nocollege6=1;
else if dmdhsedu=4 OR dmdhsedu=5 then nocollege6=0;
else if dmdhsedu IN (7,9,.) then nocollege6=.;
```

Always Check Your Work

Regardless of the recoding technique used, the process for checking your work is the same. For a categorical variable, use PROC FREQ and cross tabulate the old and new variables with the /MISSING or /MISSPRINT option.

Summary

There are a variety of ways to recode data, including the use of OR, IN( ), BETWEEN, and Elimination. The simple recoding we used in Chapter 8 will always work, but the methods covered in this lesson may be quicker and easier. Recoding methods can be combined at the discretion of the programmer. Regardless of the recoding technique, you should always check your work using PROC FREQ with the /missing option.

Lesson #10—Study Questions

1. What is the OR recoding strategy, and how does it differ from simple recoding.
2. Explain the logic behind this code:

```
if 1 le dmdhsedu le 3 then nocollege6=1;
else if dmdhsedu=4 OR dmdhsedu=5 then nocollege6=0;
else if dmdhsedu IN (7,9,.) then nocollege6=.;
```

3. How can you use the mathematical operator IN( ) to recode variables?
4. What does it mean to recode by elimination?
5. How can you combine recoding techniques?
6. What strategy should you use to check your recoding work?

Lesson 11

Recoding Continuous Variables

Learning Outcomes—Students Will:

▶ Demonstrate collapsing continuous variables into categories, and

▶ Demonstrate using continuous variables to perform calculations in SAS.

In Lesson 8, we used IF-THEN logic to clean out missing and invalid values from continuous variables. In Lessons 9 and 10, we learned how to use multiple recoding techniques to reduce the number of categories in categorical data. In this lesson, we will use these same skills to transform continuous variables into categories and to perform calculations using continuous variables.

Recall from our previous lessons that the first step is to understand the data we are working with. We will use the data dictionary, PROC CONTENTS, PROC MEANS, and/or PROC UNIVARIATE to determine how dirty the data are, how the missing/invalid values are denoted, and to make decisions about how we'd like to recode the data. Because we are dealing with continuous variables, PROC FREQ is usually not the most useful tool.

Reviewing Data Cleaning for Continuous Variables

Let's begin by reviewing how we cleaned out missing and invalid data in Lesson 8 by looking at the variable MCQ150Q (number of school days missed because of injury or illness) in the NHANES Medical Conditions DataSet. From the codebook, we learn that there is a range of valid values from 0 days to 180 days. There are also two invalid values, refused = 77777 and don't know = 99999. Only two persons responded don't know and zero responded refused. An additional 6,612 are missing. From the target, we learn this high number of missing is due to the age restrictions allowing only those 6 to 19 years old to be asked this question. Further investigation reveals that there is a screener question MCQ150G that precedes question MCQ150Q in which those 6–19 years old who did not attend school are excluded.

MCQ150Q		Target	
		B(6 Yrs. to 19 Yrs.)	
Hard Edits		**SAS Label**	
0 to 366		# school days missed from injury/illness	
English Text: During the past 12 months, that is, since {DISPLAY CURRENT MONTH} of {DISPLAY LAST YEAR}, about how many days did {you/SP} miss school because of an illness or injury?			
English Instructions: IF NONE, ENTER 0. ENTER NUMBER OF DAYS.			

Code or Value	Description	Count	Cumulative	Skip to Item
0 to 180	Range of Values	3208	3208	
77777	Refused	0	3208	
99999	Don't know	2	3210	
.	Missing	6612	9822	

Figure 11.1

We decide that we will remove the 77777 and 99999, and leave everything else the same. Within a DATA step we use the following code:

```
if mcq150q=99999 then daysmissed=.;
else if mcq150q=77777 then daysmissed=.;
else if mcq150q=. then daysmissed=.;
else daysmissed=mcq150q;
```

This code makes a new variable called daysmissed, removes the 99999 and 77777, and leaves everything else the same. Following the recode, I run PROC MEANS to check my work as follows:

```
proc means data=eight;
var mcq150q daysmissed;
run;
```

Resulting in the following:

Variable	Label	N	Mean	Std Dev	Minimum	Maximum
MCQ150Q	# school days missed from injury/illness	3210	65.8199377	2495.61	0	99999.00
daysmissed		3208	3.5174564	7.5124058	0	180.0000000

Figure 11.2

In the original variable, MCQ150Q, we can see that the mean number of school days missed because of illness or injury was 65.8 days—an awfully high number! After removing the refused and don't know, we see that the new variable has a mean of 3.5 days of school missed, a much more reasonable and plausible average. We also see that the maximum has decreased from 99999 to 180. As 180 is a value within the valid range for the data, we consider this data to be clean and can move on to our analysis.

Recoding Continuous Variables into Categories

Let's look at another example. This time, we will clean out the missing and invalid values but also recode the continuous variable into categories.

This variable DMDHRAGE is the age of the Household Reference Person (HHRP). This variable has been coded as partially continuous and partially categorical. As you can see, the variable has a valid range of 17 to 84 years and has been topcoded at 85 years. This means that any HHRP 85 years or older was coded as 85. Topcoding is typically done to preserve anonymity in national surveys. As you can see, very few heads of household (n = 163) are 85 years or older, thus reporting this in actual ages might increase the possibility of identifying a survey respondent.

DMDHRAGE	Target			
	B(0 Yrs. to 150 Yrs.)			
Hard Edits	**SAS Label**			
	HH Ref Person Age			
English Text: Age in years of the household reference person at the time of HH screening. Individuals 85 and over are topcoded at 85 years of age.				
English Instructions:				
Code or Value	**Description**	**Count**	**Cumulative**	**Skip to Item**
17 to 84	Range of Values	10185	10185	
85	>= 85 years of age	163	10348	
.	Missing	0	10348	

Figure 11.3

For this variable, we will want to keep all the valid ages of 17 to 85 years. No one is missing in this case, so we'll proceed with making categories. First, we must decide how we want to collapse the categories. This is usually done through a combination of conceptual definitions and limitations of the data. In this case, we decide that we want to classify them by decades.

We'll do <20 years 20–29 years, 30–39 years, 40–49 years, 50–59 years, 60–69 years, 70–79 years, and 80+ years.

Table 11.1

Value of New Variable	Value(s) of Old Variables
1 = <20 years	17, 18, 19
2 = 20–29 years	20, 21, 22, 23, 24, 25, 26, 27, 28, 29
3 = 30–39 years	30, 31, 32, 33, 34, 35, 36, 37, 38, 39
4 = 40–49 years	40, 41, 42, 43, 44, 45, 46, 47, 48, 49
5 = 50–59 years	50, 51, 52, 53, 54, 55, 56, 57, 58, 59
6 = 60–60 years	60, 61, 62, 63, 64, 65, 66, 67, 68, 69
7 = 70–79 years	70, 71, 72, 73, 74, 75, 76, 77, 78, 79
8 = 80+ years	80, 81, 82, 83, 84, 85

Any of the recoding techniques covered in the previous lesson will work; however, most of them will be very inefficient. The "between" strategy is the most useful when working with continuous variables. Within the DATA step, we use the following code:

```
data recode1; set nhanes.demo_d;
if 17 le dmdhrage le 19 then hhrpage=1;
else if 20 le dmdhrage le 29 then hhrpage=2;
else if 30 le dmdhrage le 39 then hhrpage=3;
else if 40 le dmdhrage le 49 then hhrpage=4;
else if 50 le dmdhrage le 59 then hhrpage=5;
else if 60 le dmdhrage le 69 then hhrpage=6;
else if 70 le dmdhrage le 79 then hhrpage=7;
else if 80 le dmdhrage le 85 then hhrpage=8;
else hhrpage=.;
run;
```

Since my final variable (HHRPAGE) is categorical, I will use PROC FREQ to check my work. Again, for ease of display, I am suppressing the column, row, and total percentages.

```
proc freq data=recode1;
tables dmdhrage*hhrpage /missing nocol norow nopercent;
run;
```

This is a partial table of the output from the earlier frequency procedure:

Frequency	Table of DMDHRAGE by HHRPage									
DMDHRAGE(HH Ref Person Age)	HHRPage									
	1	2	3	4	5	6	7	8	Total
17	3	0	0	0	0	0	0	0	3
18	93	0	0	0	0	0	0	0	93
19	30	0	0	0	0	0	0	0	130
20	0	116	0	0	0	0	0	0	116
21	0	119	0	0	0	0	0	0	119
22	0	125	0	0	0	0	0	0	125
23	0	130	0	0	0	0	0	0	130
24	0	198	0	0	0	0	0	0	198
25	0	221	0	0	0	0	0	0	221
26	0	237	0	0	0	0	0	0	237
27	0	234	0	0	0	0	0	0	234
28	0	213	0	0	0	0	0	0	213
29	0	234	0	0	0	0	0	0	234
30	0	0	202	0	0	0	0	0	202

Figure 11.4   As before, I am checking to see that the values of the old variable that I wanted to combine are all present in the values of the new variable. In this example, 17, 18, and 19 are all now coded as 1; 20–29 are now all coded as 2, and so on. You would continue this process to check each of the eight values of the new variable.

Finally, I run a PROC FREQ on the newly recoded variable:

hhrpage	Frequency	Percent	Cumulative Frequency	Cumulative Percent
1	226	2.18	226	2.18
2	1827	17.66	2053	19.84
3	2682	25.92	4735	45.76
4	2557	24.71	7292	70.47
5	1218	11.77	8510	82.24
6	897	8.67	9407	90.91
7	566	5.47	9973	96.38
8	375	3.62	10348	100.00

Figure 11.5

Here I can see that instead of 68 values, I have only 8.

Calculations with Continuous Variables

Another helpful recoding strategy is allowing SAS to calculate variables from the values of other variables. For example, perhaps I wanted to examine BMI, and I had variables for height in meters and weight in kilograms, but I did not have a variable for BMI in my data. I know that BMI is the weight in kilograms divided by the height in meters squared. I could create the variable BMI using the following syntax within a DATA step:

```
BMI=weightkg/heightmeters**2;
```

Where WEIGHTKG is the variable containing the values for weight in kilograms, HEIGHTMETERS is the variable containing the values for height in meters, and **2 tells SAS to square the results.

SAS will conduct nearly any calculation that you can think of. SAS follows the standard order of operations (parentheses, exponents, multiplication, division, addition, and subtraction).

The following are some examples of simple calculations in SAS:

Table 11.2

Calculation	Syntax
Multiplication	newvar=var1*var2
Division	newvar=var1/var2
Addition	newvar=var1+var2
Subtraction	newvar=var1-var2
Exponents[§]	newvar=var1**2

§ The two asterisks indicate an exponent and the 2 = squared, **3 = cubed, etc.

I have given all of these examples using two variables (except exponents), but values can also be used. For example, if I wanted to convert age in months to age in years, I could use the following code:

```
ageinyears=ageinmonths/12;
```

Where AGEINYEARS and AGEINMONTHS each represent a variable containing that data.

Let's look at an example. NHANES collects data on a variety of laboratory measures. Suppose that we are interested in determining the percent of total cholesterol that is HDL cholesterol. Total cholesterol is reported in a dataset called TCHOL_D, and HDL cholesterol is reported in a dataset called HDL_D. To perform a calculation on these variables, I first merged these two datasets together, resulting in a new dataset I've called CHOL with 8,086 observations and 5 variables. Now I can look at the data dictionary for each of the original datasets (TCHOL_D & HDL_D) to determine which variables I should use.

From the TCHOL_D dataset, I see that total cholesterol in mg/dL is represented in the variable LBXTC. Here we can see that there are 7,360 valid observations and that the variable is continuous ranging from 78 to 615. There are missing data, but no invalid data, so I do not need to clean the data prior to my calculation.

LBXTC - Total cholesterol (mg/dL)

Variable Name:	LBXTC
SAS Label:	Total cholesterol (mg/dL)
English Text:	Total cholesterol (mg/dL)
Target:	Both males and females 6 YEARS - 150 YEARS

Code or Value	Value Description	Count	Cumulative	Skip to Item
78 to 615	Range of Values	7360	7360	
.	Missing	726	8086	

Figure 11.6

Similarly, I see that HDL cholesterol is represented by the variable LBDHDD.

LBDHDD - Direct HDL-Cholesterol (mg/dL)

Variable Name:	LBDHDD
SAS Label:	Direct HDL-Cholesterol (mg/dL)
English Text:	Direct HDL-Cholesterol (mg/dL)
Target:	Both males and females 6 YEARS - 150 YEARS

Code or Value	Value Description	Count	Cumulative	Skip to Item
15 to 188	Range of Values	7360	7360	
.	Missing	726	8086	

Figure 11.7

I would like to determine the percentage of total cholesterol that is HDL cholesterol. To do this, I need to divide HDL cholesterol by total cholesterol. In a standard mathematics equation, this would look like this:

$$\frac{HDL\ Cholesterol}{Total\ Cholesterol} \times 100.$$

In SAS, I would use the following syntax:

```
data crecode; set chol;
percenthdl = (LBDHDD/LBXTC)*100;
run;
```

Note that as with all other recoding steps, this is done within the context of a DATA step. This can be combined with multiple recoding steps, as we have done previously.

As always, I now need to check my work to make sure I'm getting the results I expect. To do this, I will run a three-way cross tabulation using PROC FREQ. For this example, we have three variables: LBDHDD, LBXTC, and my newly recoded PERCENTHDL. To check my work here, I need to see all three variables at the same time. Prior to this, we have always looked at cross tabulations of two variables. By default, adding a third dimension (third variable) to the tables produces a set of row*column tables stratified by the third variable. In this example, that would produce a large number of tables, and would complicate checking our work. When checking recodes that are a combination of three or more variables, the LIST option displays a much more concise table.

```
proc freq data = crecode;
table lbdhdd*lbxtc*percenthdl/list missprint;
run;
```

As we did previously in the TABLES statement, I am separating my variables by asterisks. This tells SAS that I would like to cross tabulate. Since we wish to cross tabulate three variables (instead of the normal 2), we'll need to use the LIST option in SAS. Notice that like MISSING or MISSPRINT, which we have previously used, LIST appears in the TABLES statement of PROC FREQ, following the list of variables and a forward slash. Using the LIST option produces the following output:

(Partial Table)

LBDHDD	LBXTC	percenthdl	Frequency	Percent	Cumulative Frequency	Cumulative Percent
.	.	.	726	.	.	.
15	145	10.344827586	1	0.01	1	0.01
17	128	13.28125	1	0.01	2	0.03
17	269	6.3197026022	1	0.01	3	0.04
20	120	16.666666667	1	0.01	4	0.05
20	135	14.814814815	1	0.01	5	0.07
21	106	19.811320755	1	0.01	6	0.08
21	182	11.538461538	1	0.01	7	0.10
22	143	15.384615385	1	0.01	8	0.11

Figure 11.8   Notice that for observations that were missing data on both variables, the result of the calculation is missing, not zero. If data is missing on any variable involved in the calculation, the result will be missing.

Notice that the variables are ordered as they were in my TABLES statement. I intentionally chose this order since it follows the natural flow of the calculation. Depending on the type of calculation, you may prefer a different order.

I can check my work by looking across the rows to see if the values are what I would expect. Let's look at the first row; this respondent has HDL cholesterol of 15 and total cholesterol of 145. If I manually perform the calculation ((15/145)*100 = 10.3448), I get the same answer SAS has calculated for me. It's always a good idea to spot-check and be sure that the calculation is doing what you think it should be doing.

I can also check my work using PROC PRINT. PROC PRINT has some advantages over PROC FREQ. In this case, it primarily allows me to restrict the number of printed observations, thus reducing the size of the table.

```
proc print data = crecode (obs=5);
var lbdhdd lbxtc percenthdl;
run;
```

Using PROC PRINT with the VAR statement and the (OBS = xx) option produces the following output. As with the PROC FREQ output, I can manually spot-check my calculation to see that it is working as expected. For example, observation 1 has an HDL cholesterol of 55 and a total cholesterol of 129: (55/129)*100 = 42.6357.

Obs	LBDHDD	LBXTC	percenthdl
1	55	129	42.6357
2	46	170	27.0588
3	.	.	.
4	39	105	37.1429
5	59	147	40.1361

Figure 11.9

Additional Options for Calculations

In the previous output tables, you likely noticed that SAS is calculating percenthdl out to four or more decimal places (depending on the type of output). In practice, we may wish to report out to only a couple of decimal places or perhaps as whole numbers.

SAS will conduct a variety of numeric manipulations. This is a partial list of keywords that tell SAS to manipulate numeric data, including an example of how using each function would change a variable's values.

Table 11.3

Function	Syntax	Old Value	New Value
Absolute Value	newvar=abs(oldvar)	−4	4
Round to Nearest Whole #	newvar=round(oldvar, 1)	24.573	25
Round to Nearest Tenth	newvar=round(oldvar, 0.1)	24.573	24.6
Integer	newvar=int(oldvar)	24.573	24
Square Root	newvar=sqrt(oldvar)	81	9
Natural Log	newvar=log(oldvar)	24.573	3.2016

For example, suppose I wish to report percenthdl as one decimal place only. To do so, I could create a new variable, where percenthdl is rounded as follows:

```
data crecode; set chol;
percenthdl = (LBDHDD/LBXTC)*100;
percenthdl2 = round(percenthdl,0.1);
run;
```

Running PROC PRINT again, I can see that my new variable percenthdl2 is rounded to the nearest tenth.

Obs	LBDHDD	LBXTC	percenthdl2
1	55	129	42.6
2	46	170	27.1
3	.	.	.
4	39	105	37.1
5	59	147	40.1

Figure 11.10

If I decide that I would like to categorize my percent HDL variable, I could recode as we did in the previous step. This could be done in the same DATA step, so long as the calculation occurs before any subsequent recoding. For example, if I wish to recode my variable percenthdl2 into a dichotomous variable where 1 = Percent HDL ≥30 and 0 = Percent HDL <30, I could use the following syntax:

```
data crecode; set chol;
percenthdl = (lbdhdd/lbxtc)*100;
percenthdl2 = round(percenthdl,0.1);

if percenthdl2 ge 30 then hdlperc = 1;
else if 0 le percenthdl2 lt 30 then hdlperc = 0;
run;
```

If I run PROC FREQ on the HDLPERC variable, I get the following table.

hdlperc	Frequency	Percent	Cumulative Frequency	Cumulative Percent
0	3645	49.52	3645	49.52
1	3715	50.48	7360	100.00
Frequency Missing = 726				

Figure 11.11

Now I will illustrate one of the common pitfalls of recoding continuous variables, particularly using the between strategy. Given that my objective in recoding my HDL percent variable into a dichotomous variable is to make everything ≥30 have a value of 1 and everything < 30 have a value of 0, it seems reasonable that I could use the following recode. Note that my else-if statement reads: else if percenthdl2 lt 30 then hdlperc = 0;. This seems intuitive and perfectly reasonable. Let's run this code and see what happens.

```
data crecode; set chol;
percenthdl = (LBDHDD/LBXTC)*100;
percenthdl2 = round(percenthdl,0.1);

if percenthdl2 ge 30 then hdlperc = 1;
else if percenthdl2 lt 30 then hdlperc = 0;
run;
```

I output the following table using a PROC FREQ on my newly recoded variable. Compared to the preceding table, I have the same number of 1s (3,715), but a different number of 0s (4,371 vs. 3,645).

hdlperc	Frequency	Percent	Cumulative Frequency	Cumulative Percent
0	4371	54.06	4371	54.06
1	3715	45.94	8086	100.00

Figure 11.12

I need to check my work to find out what is different. As always, I'll do this using a cross tabulation of my old and new variables.

```
proc freq data = crecode;
tables percenthdl*hdlperc/missprint norow nocol nopercent;
run;
```

(Partial Table)

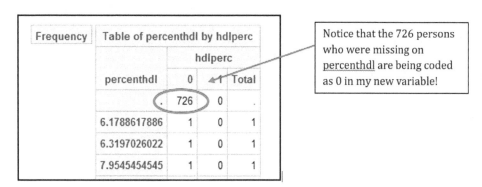

Figure 11.13

In the earlier table, you can see that using the less than (lt) or less than or equal to (le) syntax can have unintended consequences. SAS treats missing (.) as a value less than 0 in coding logic. Therefore, in this situation where I have specified else if percenthdl2 lt 30 then hdlperc = 0, all of my missings are coded as zero.

This example underscores the importance of checking your recodes using PROC FREQ or PROC PRINT. Looking at the code alone, I would have completely missed this error and falsely reported my data. I can easily correct this mistake by putting a lower bound on the range in my code.

Summary

Cleaning and recoding continuous variables into categories follows the same rules and uses the same techniques previously covered with categorical variables. SAS can use continuous variables to perform a variety of mathematical calculations. This is a useful tool for calculating data that you want but which are not readily given to you. The LIST option in PROC FREQ allows you to easily check calculations that involve three or more variables.

Lesson #11—Study Questions

1. What does it mean when a variable is topcoded?
2. How do you check your work when converting continuous variables into categories?
3. Why should you check your work when recoding variables?
4. What is the most efficient strategy for recoding continuous variables into categories?
5. What is the BETWEEN syntax, and how does it work?

6. If your dataset included the amount of time it took students to take a test in hours (var=time), what syntax would you use to convert it to minutes?

7. Fill in the syntax for the following calculations and functions:

Table 11.4

Function	Syntax
Natural log	
Round to the nearest whole number	
Exponents	
Addition	
Absolute value	
Multiplication	
Round to the nearest tenth	
Integer	
Square root	
Division	
Subtraction	

Lesson 12

Comments, Labels, and Titles

Learning Outcomes—Students Will:

▶ Demonstrate using comments to annotate SAS programs,

▶ Apply labels to variables to improve SAS output, and

▶ Use titles to add descriptive information to SAS output.

Up to this point, we have learned to open and import datasets in SAS and to manipulate variables. The next two lessons will deal with ways to improve the readability and organization of your SAS programs (comments) and the resulting output (labels, titles, formats). These skills will significantly improve the overall quality of your coding and make the programs and output more useful to you and others involved in the project.

Comments, labels, and titles are all methods of documenting your SAS coding and output. Documentation of your decision making and methods in the SAS program are extremely important skills for a well-rounded SAS programmer. Documentation relieves you of the burden of having to remember the rationale for every decision you make and allows others who use your program to, if necessary, pick up where you left off. Using these concepts will enhance the usability of your syntax and output.

Comments

Comments help you document your decision making within your SAS program, help you later recall the decisions you made regarding cleaning and recoding, help you remind yourself of procedures, and allow you to make notes about the data and/or your findings. Comments are visible in your SAS program as green text, but they are ignored by SAS when the program is run. These are notes just for you and other persons who may use your program.

SAS comments can be created using one of two methods in SAS:
Method #1 = /* Commented Text */
Method #2 = * Commented Text;

In the first method, you tell SAS that you are beginning a comment using a forward slash followed by an asterisk, as follows: (/*). Following this, all text or symbols entered will be considered by SAS to be comments until you close the comment string using an asterisk followed by a forward slash: (*/). In the second method, an asterisk opens the comment (*) and a semicolon (;) closes the comment. Regardless of the method, SAS treats all comments the same. The only difference is the *; comment does not work in the middle of a SAS statement, whereas the /* */ comment indicators will ALWAYS work. This method (method 1) allows you to comment across multiple SAS statements and therefore can be useful for deactivating statements you wish to temporarily remove from your program. The following, for example:

```
data chol; merge bios.hdl_d bios.tchol_d;
by seqn;
run;
/*
data crecode; set chol;
percenthdl = (lbdhdd/lbxtc)*100;
percenthdl2 = round(percenthdl,0.1);

if percenthdl2 ge 30 then hdlperc = 1;
else if percenthdl2 lt 30 then hdlperc = 0;
run;
*/
data crecode; set chol;
percenthdl = (lbdhdd/lbxtc)*100;
percenthdl2 = round(percenthdl,0.1);

if percenthdl2 ge 30 then hdlperc = 1;
else if 0 le percenthdl2 lt 30 then hdlperc =
0;
run;
```

When I run this program, SAS will ignore the data step that is commented out.

Figure 12.1

What Should I Comment?

What you comment is up to you, but some suggested items include

- who wrote the program,
- when it was written,
- why it was written,
- data cleaning and recoding decisions,
- steps taken in the analysis,
- reminders or "notes to self" for when you come back to a program later, and
- notes on the results of the program.

Here are some examples of comments that might be useful:

1. This is commented text from a SAS program I wrote for a manuscript. This tells me what project this program belongs to, and I've indicated it's a recode file. You will note that you can draw lines in comments or other things to delineate breaks or sections in your code; this can be helpful but is not necessary.

```
/************************************************************

    Age at first intercourse and sexual partnering among
    female respondents to the NSFG


    Magnusson BM, Nield J, Lapane KL
    Virginia Commonwealth University
    Department of Epidemiology & Community Health


    Recode File
************************************************************/
```

2. This is a section of data recodes from the same project. Here you can see that I delineate the section of recodes. The sociodemographic variables and the individual variables I am looking at include age and race/ethnicity. Further, following each statement, I note what that category represents.

```
*********************************************************
Sociodemographic Variables
*********************************************************;
/*age*/
if 21 le ager le 24 then age=1; *15-19 years;
else if 25 le ager le 29 then age=2; *25-29 years;
else if 30 le ager le 34 then age=3; *30-34 years;
else if 35 le ager le 39 then age=4; *35-39 years;
else if ager ge 40 then age=5; *40-44 yrs;


/*race and ethnicity*/
if hisprace=2 then race=1;*white;
else if hisprace=3 then race=2;*black;
else if hisprace=1 then race=3;*Hispanic;
else if hisprace=4 then race=4;*other;
else race=.;
```

3. This section illustrates leaving notes that may be important in helping me recall the methodology. In this section, I am undertaking a complex recode of a variable to define months at risk of pregnancy in the 12 months prior to interview. Because the interviews happened across a period of several years, the 12 months prior to the interview are not constant for all women (e.g., one woman may be June to May and another February to January). These notes help me determine how to handle the recode and, when I come back later, remind me what I did/was trying to do.

```
********************************************************
*    Define contraceptive use
*    1) Identify the period of interest
     (12 months prior to interview)
*    2) Collapse across 4 mentions with some defined
     hierarchy
*    3) Count the number of months of contraceptive
     use;
********************************************************;
*NOTE;/*In 06-08 data the methods used are collected from Janu-
ary[intyr-3] to December[intyr] therefore respondents interviewed in
December 06 were evaluated from Jan 03-Dec 06 and respondents in-
terviewed in June 07 were evaluated from Jan 04-Jun 07. This vari-
able pattern requires isolation of the interview month and year and
counting back from that point to identify the 12 months prior to the
interview.*/
```

Comments can dramatically increase the usability of your program and should be used liberally.

Labels

You have likely noticed that many of the variables we have used up to this point have nonsensical or cryptic names. They are typically some combination of letters and numbers, and they may have some meaning, but that meaning may not be particularly intuitive. Labels are used to translate these cryptic variable names into something more descriptive and meaningful. The label appears only in the output and improves the readability of your output. Within your syntax, you must still refer to your variable by its name. Labels can be applied in a DATA step or in a PROC step.

Creating Labels in a DATA Step

Labels applied in the DATA step are permanently associated with the variable and will remain associated with the variable in every procedure you run, in subsequent copies of the dataset you make, etc. This is typically the preferred method since labels applied in the PROC step are temporary and will appear only in the output from that specific PROC step.

The syntax is as follows:

```
data newdatasetname; set olddatasetname;
/*Recoding Steps*/
label varname='Label that you want to apply';
run;
```

Typically, you will apply your labels in the same DATA step in which you do your recoding. There are no specific rules about this, except that the label cannot be created until the variable has been created. For this reason, you often see label statements near the bottom of the DATA step.

The SAS keyword is LABEL followed by the variable name for the variable you are labeling, an equal sign (=), and the text you wish to display as the label within single or double quotes. SAS is not picky about single or double quotes so long as you open and close the quoted text using the same type of quotation marks. The exception to this rule is if you have a label that contains a word with an apostrophe. (e.g., Respondent's Education). The apostrophe will be read as a quotation mark. Therefore, labels that contain apostrophes should be opened and closed with double quotation marks. The label will apply exactly as you type it, including spelling, capitalization, and punctuation.

The following is a frequency table of the recode we did in Lesson 3. The variable name is nocollege1:

nocollege1	Frequency	Percent	Cumulative Frequency	Cumulative Percent
0	2771	50.11	2771	50.11
1	2759	49.89	5530	100.00

Frequency Missing = 4818

Figure 12.2

Within the same DATA step in which I created the variable, I can also label it. At any point in the program after the variable has been created, I can add the following SAS statement:

```
label nocollege1='No college education';
```

If I again run a frequency of this variable, I can see that in addition to the variable name (nocollege1), the label has been added as "No College Education":

No college education				
nocollege1	Frequency	Percent	Cumulative Frequency	Cumulative Percent
0	2771	50.11	2771	50.11
1	2759	49.89	5530	100.00

Frequency Missing = 4818

Figure 12.3

A DATA step in which labels and recodes are completed might look like this:

```
Data recode; set nhanes.meddemo;
if dmdhsedu in (1,2,3) then nocollege3 = 1;*No college education;
else if dmdhsedu in (4,5) then nocollege3 = 0; *At least
some college;
else if dmdhsedu in (7,9,.) then nocollege3 = .;

if mcq150q=99999 then daysmissed=.; *Set DK/R to missing;
else if mcq150q=77777 then daysmissed=.;
else if mcq150q=. then daysmissed=.;
else daysmissed=mcq150q;

label nocollege3 = 'No college education' daysmissed = 'No. School
Days missed for illness/injury';
run;
```

Note that I am using one label statement for all of the variables I am recoding. The keyword LABEL only appears once followed by the syntax for the variable name = 'label'; it is repeated for as many variables as you are labeling (here are two variables: (1) nocollege3 and (2) days missed. The label statement ends with a semicolon when all variables have been labeled.

Creating Labels in a PROC Step

Typically, it is preferred to apply the labels in a DATA step so that they will be permanently associated with the variable. However, in some cases, it may be desirable to quickly add a label in a PROC step. Adding a label in a PROC step is faster, but it's also temporary and will only appear in the output for that particular PROC step.

For this example, we will use PROC FREQ, although the same applies in any procedure you use.

```
proc freq data=class.dataset;
tables variable2;
label variable2='recoded age groups';
run;
```

Note that the syntax is exactly the same: LABEL variable name = 'Label'; . When labeling in a PROC step, you may label any variable or variables that is/are found in the TABLES or VAR statement of the procedure.

Titles

Titles are used to document output. They are not applied to a particular variable or value but appear at the beginning of the output.

By default, SAS prints the titles *The SAS System* followed by the name of the procedure you ran such as *The FREQ procedure* or *The MEANS procedure*, as shown next:

The SAS System

The FREQ Procedure

No college education

nocollege1	Frequency	Percent	Cumulative Frequency	Cumulative Percent
0	2771	50.11	2771	50.11
1	2759	49.89	5530	100.00

Frequency Missing = 4818

Figure 12.4

Using a title statement, you can replace these titles with ones that are more descriptive. Titles <u>may be applied anywhere in a SAS statement and anywhere in a SAS program</u>. Much like a libname statement, the title statement does not need to be contained within a DATA step or a PROC step.

The syntax for a title statement is TITLE followed by whatever you want printed in single or double quotation marks. As with labels, if your title contains an apostrophe, you must use double quotation marks. In this example, you will note that the title statement comes prior to the DATA step. However, it may be within the DATA step or within a PROC step.

```
Title 'Household Reference Education Recoding';
data one; set nhanes.demo_d;
if dmdhsedu in (1,2,3) then nocollege1=1;
else if dmdhsedu in (4,5) then nocollege1=0;
else if dmdhsedu in (7,9,.) then nocollege1=.;
label nocollege1='No college education';
run;

proc freq data=one;
tables nocollege1;
run;
```

Running the previous code produces this output, which you can see replaces *The SAS System* with the title I have requested: *Household Reference Education Recoding*.

Household Reference Education Recoding

The FREQ Procedure

	No college education			
nocollege1	Frequency	Percent	Cumulative Frequency	Cumulative Percent
0	2771	50.11	2771	50.11
1	2759	49.89	5530	100.00

Frequency Missing = 4818

Figure 12.5

Regardless of where I put the title statement, the syntax and the result will be the same.

Within a DATA Step

```
data one; set nhanes.demo_d;
if dmdhsedu in (1,2,3) then nocollege1=1;
else if dmdhsedu in (4,5) then nocollege1=0;
else if dmdhsedu in (7,9,.) then nocollege1=.;
label nocollege1='No college education';
Title 'Household Reference Education Recoding';
run;
```

Within a PROC Step

```
proc freq data=one;
tables nocollege1;
Title 'Household Reference Education Recoding';
run;
```

Multiple Levels of Titles

Using the keywords TITLE1, TITLE2, TITLE3, etc., will allow you to create multiple levels of titles. For instance, I may wish to give my project a global title, which I will use on all output for the project, I might use TITLE1 and then give each procedure its own title using TITLE2. You can add as many levels of titles as desired, but in general, you should be judicious, as the titles will print on every page of output.

Here I am adding the global title (TITLE1): SAS Computing—Titles Demonstration and titling each of my PROC steps separately using TITLE2. The Frequency Procedure is given the title Raw Frequencies and the Contents Procedure is titled Contents for DataSet RECODE.

```
title1 'SAS Computing - Titles Demonstration';
proc freq data=recode;
tables unhealthy worse noplace placetype;
title2 'Raw Frequencies ';
run;

proc contents data=recode;
title2 'Contents for Data Set RECODE';
run;
```

After running this code, you can see that these titles now appear in my SAS output:

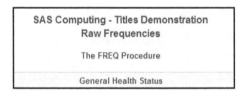

Figure 12.6

Removing Titles

Because titles are not restricted to a dataset or procedure, they will continue to appear in all output during your SAS session until you (1) overwrite them with a new title or (2) clear the title.

For example, this PROC step is run following the contents statement, but it does not include a new title. As you can see, the title still says this is the contents procedure, which is obviously incorrect.

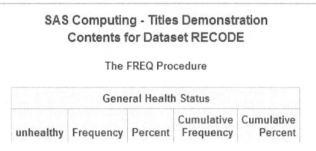

Figure 12.7

Overwriting Titles

To overwrite the title, use the same keyword (e.g., TITLE, TITLE1, TITLE2) and new text. In the example from earlier, I used TITLE2 as the keyword in both PROC steps. Using TITLE2 in the PROC CONTENTS overwrites the previous title.

```
title1 'SAS Computing - Titles Demonstration';
proc freq data=recode;
tables unhealthy worse noplace placetype;
title2 'Raw Frequencies ';
run;

proc contents data=recode;
title2 'Contents for Data Set RECODE';
run;
```

Clearing Titles

To clear a title, you need to run an empty title statement. As with other title statements, this may appear anywhere in the SAS program. An empty title statement consists of only the keyword for the title you want to clear and a semicolon.

```
title2;
```

This clears title2, but title1 will remain, as shown here:

```
title2;
proc freq data=recode;
tables unhealthy;
run;
```

Which results in the following:

SAS Computing - Titles Demonstration

The FREQ Procedure

General Health Status				
unhealthy	Frequency	Percent	Cumulative Frequency	Cumulative Percent

Figure 12.8 Note that the TITLE1 remains and TITLE2 has been cleared. I would use a similar procedure with keyword TITLE1 to clear the first title.

Summary

Comments, labels, and titles are all methods of documenting your SAS coding and output.

Comments, which appear in green, have several purposes, including helping you recall decisions you made about cleaning and recoding, helping you remind yourself of procedures, and allowing you to make notes about the data and/or findings.

Labels allow you to attach descriptive and meaningful names to your variables. These labels appear in your output and can be applied in either a DATA step or a PROC step. It is important to note that if applied in a PROC step, the label will only appear in the output from that specific PROC step.

Titles allow you to attach descriptive and meaningful titles to the procedures you run. Titles can be applied before, after, or within DATA steps or PROC steps. Multiple levels of titles can be used, as needed, to provide meaning to the output that is created. Once created, titles will appear with all output until overwritten or cleared.

Using these three tools, the usefulness of syntax and output can be enhanced.

Lesson #12—Study Questions

1. What are the purposes of using comments?
2. How often should you use comments?
3. What are the two methods of commenting, and why would you use one over the other?
4. What is the purpose of using labels?
5. What is the difference between creating labels in a DATA step as compared to creating labels in a PROC step?
6. What are titles?
7. Why would you use multiple levels of titles?
8. How do you overwrite a title?
9. How do you clear a title?
10. Put a check in the boxes that apply to each of the documentation methods:

Table 12.1

	Purple Text	Green Text	Can Be Used Before DATA/ PROC Step	Can Be Used within DATA Step	Can Be Used within PROC Step
Comments					
Labels					
Titles					

Lesson 13

Formatting Variables

Learning Outcomes—Students Will:

▶ Explain the difference between permanent and temporary format libraries,

▶ Demonstrate how to create a temporary format library using the PROC FORMAT procedure,

▶ Demonstrate applying formats in a DATA step and a PROC step, and

▶ Demonstrate the use of formats to change the order of values in the output display.

Formats

The previous chapter covered comments, titles, and labels, which are all useful in improving your output. This lesson will cover creating and applying formats, another skill that can be used to make your output more user-friendly. Formats are used to describe the levels of variables called values. Typically, data are coded as numbers because numeric variables are easier to use, and recoding is simpler with numeric, rather than character, values. Yet often the numbers stand for something else. For example, I might have a variable for gender that is coded as 1 or 2 where 1= Male and 2 = Female. Formats allow me to add the descriptors "male" and "female" to my output, yet still use the faster, easier numbers behind the scenes.

Formats come from a variety of different places:

1. SAS has some standard formats for things like dates and social security numbers.
2. Many Centers for Disease Control and Prevention or NCHS datasets, such as the NHANES used in this book, have a format library provided with the data.
3. You can create your own formats.

As it is generally most helpful to have formats for variables you will be using in your analysis, we will be focusing on how to create your own formats.

Steps for Creating Formats

There are two steps for creating formats:

1. Create a format library
2. Apply the formats

Step 1: Create a Format Library

A format library is a list of formats that can be used to add labels to the values of variables (e.g., 1 = "Male"; 2 = "Female"). You create a format library using the format procedure (PROC FORMAT). Creating the format library is always the first step in applying formats. Once the format is created and stored in the format library, you can attach your formats to variables in your dataset. A format library may contain any number of formats.

Temporary versus Permanent Format Libraries

Format libraries may be either permanent or temporary. A format library is similar to a library you create to store your datasets in, except in this situation, the library contains formats instead of datasets. Temporary format libraries behave like the WORK library. They exist only during your SAS session. When SAS is closed, the library goes away. When you come back to SAS, if you wish to use the same library, your PROC FORMAT code must be re-run. There are many positives to using a temporary format library: (1) they are simpler to use; (2) they are easily modified; (3) you can still apply them permanently. However, there is one downside. As mentioned previously, the PROC FORMAT must be run each time you open SAS, and you will be unable to open a dataset with formats assigned until you run the format library or suppress format errors.

A permanent library is stored permanently in SAS and thus always available to you in SAS. The positive to this is that you don't have to re-run formats each time you open SAS. However, they are more complicated to create and <u>are available only on the computer in which you set them up</u>. Since you are unable to open the dataset without the formats, transferring datasets to a different computer is tricky.

In this book, we will always create temporary format libraries and apply them permanently.

Creating a Temporary Format Library

A temporary format library is created using PROC FORMAT. The syntax looks like this:

```
proc format;
value formatname
1='Format'
2='Format';
run;
```

Note that following the format name there is NOT a semicolon. From the keyword VALUE until you complete labels for all of your levels, there is ONE SAS statement, so NO SEMICOLON.

PROC FORMAT is the procedure name. Unlike all of the other procedures we have learned up to this point, the "data =" is NOT needed. Format libraries are created within SAS, not within a specific dataset.

VALUE is the SAS keyword to identify that you are creating a format. Within PROC FORMAT, every time SAS sees VALUE, it knows this is a new format. The format name is any name you create for your format. As with variables, formats can be named virtually anything. However, format names cannot end in a number! Many programmers name their formats the same name as their variables. This is a useful strategy *if* you are only planning to use that format for one specific variable. Other times you may have many variables that could use the same format. In these cases, more generic names may be preferred. For example, you could have several variables all coded 1 = Yes; 0 = No. A generic format will save time; therefore, I could create one format for Yes/No variables and name it yn.

Following the format name, you list the value (i.e., the level of the variable: 0, 1, 2, 3), followed by an equal sign, and, within matching single or double quotes, the label you want to apply to that level. As with labels and titles, if your format contains an apostrophe, double quotation marks must be used.

Let's look at an example. I have a variable called GENDER that is coded where 1 = Male and 2 = Female.

```
proc format;
value sexf
1='1-Male'
2='2-Female';
run;
```

Here the format is named SEXF. Note that I have included the value number in the format. This makes it easier to reference in your coding later and is always a good practice. When formats are applied to a variable, only the format (not the values) appears in the output. For this reason, I always recommend including the value number within the quotations of your format.

One format procedure can contain several formats. For example, I could continue adding to the format from earlier as follows:

```
proc format;
value sexf
1='1-Male'
2='2-Female';
value yn
1='1-Yes'
0='0-No';
value race
1='1-White, Non-Hispanic'
2='2-Black, Non-Hispanic'
3='3-Hispanic, Any Race'
4='4-Other Non-Hispanic';
run;
```

Here my formats are named SEXF, YN, and RACE. Note again that I have included the values in the quotations and that the semicolons only come at the very end of the whole format. Once you have created the format, you must highlight and run this portion of the program BEFORE (or along with) running the DATA step in which you apply the formats.

Step 2: Apply the Formats

Formats may be applied or associated with the variable permanently using a DATA step (recommended) or temporarily in a PROC step.

Applying Formats Permanently in the DATA Step

When formats are applied in the DATA step, they are permanent and thus will always be associated with the variable from that point forward. Applying formats in this manner is the most common method and is useful if you plan to use the variables in multiple analyses (PROCs).

Formats are applied in the dataset using the keyword FORMAT. Formats can only be applied in the context of the DATA step (i.e., after DATA and before RUN) or a PROC step and must come after the variable you are formatting is created. When referencing a format name anywhere other than the PROC FORMAT step, the format name *must* be followed by a period. Using a period at the end (e.g., formatname.) will turn the text teal and indicates to SAS that this is a format name, not a variable name. For example:

```
data new; set old;
format varname formatname.;
run;
```

You can apply multiple formats in the same FORMAT statement. SAS is expecting a pattern of variable name then format name. Note that I have the variable name, followed by the formatname.

```
data new; set old;
format varname formatname. varname formatname. varname formatname.;
run;
```

You can also apply the same format to multiple variables as follows:

```
data new; set old;
format varname varname varname varname formatname.;
run;
```

Here, SAS will apply the same format to all four variables.

Ordering the steps of recoding and formatting is important for efficient programming. Although you may create your formats after or along with your recoding, for the most efficient program, the formats should precede the DATA step within your SAS editor window. Your recoding and formatting steps should appear in the SAS program in this order:

1. PROC FORMAT for creating your formats
2. Recoding of variables within the DATA step
3. Applying formats with the format statement inside a DATA step

Let's walk through an example from start to finish. We will recode, label, and format three different variables from the Food Security NHANES dataset.

The entries from the data dictionary are as follows:

Variable FSD032A:

FSD032A - HH Worried run out of food

Variable Name:	FSD032A
SAS Label:	HH Worried run out of food
English Text:	Now I am going to read you several statements that people have made about their food situation. For these statements, please tell me whether the statement was often true, sometimes true, or never true for {you/your household} in the last 12 months, that is since last {DISPLAY CURRENT MONTH}. The first statement is . . . {I/we} worried whether {my/our} food would run out before {I/we} got money to buy more.
English Instructions:	Collected on household level.
Target Gender:	Both males and females
Target Age:	0 YEARS - 150 YEARS

Code or Value	Value Description	Count	Cumulative	Skip to Item
1	Often true	793	793	
2	Sometimes true	1860	2653	
3	Never true	7565	10218	
7	Refused	7	10225	
9	Don't know	0	10225	
.	Missing	123	10348	

Figure 13.1

Variable FSD032B:

FSD032B - HH Food didn't last

Variable Name:	FSD032B
SAS Label:	HH Food didn't last
English Text:	[The next statement is . . .] The food that {I/we} bought just didn't last, and {I/we} didn't have money to get more.
English Instructions:	Collected on household level.
Target Gender:	Both males and females
Target Age:	0 YEARS - 150 YEARS

Code or Value	Value Description	Count	Cumulative	Skip to Item
1	Often true	503	503	
2	Sometimes true	1652	2155	
3	Never true	8063	10218	
7	Refused	7	10225	
9	Don't know	0	10225	
.	Missing	123	10348	

Figure 13.2

Variable FSD190:

FSD190 - Fd Stmp SP mos authorized in last 12 mos

Variable Name:	FSD190
SAS Label:	Fd Stmp SP mos authorized in last 12 mos
English Text:	In the last 12 months, about how many months {were you/was SP} authorized to receive Food Stamps?
English Instructions:	Collected at individual level in the household interview.
Target Gender:	Both males and females
Target Age:	0 YEARS - 150 YEARS
Hard Edits:	1 to 12

Code or Value	Value Description	Count	Cumulative	Skip to Item
2 to 12	Range of Values	872	872	
1	1 month or less	59	931	
77	Refused	1	932	
99	Don't know	6	938	
.	Missing	9410	10348	

Figure 13.3

We will recode FSD032A and FSD032B into Yes/No variables where Yes = Ever True, No = Never True. We will set all refused, don't know, and missing values to the system missing (.). We will recode FSD190 as <3 months, 3–6 months, and >6 months.

First, I start by recoding and checking my recodes using the following SAS code:

```
data foodsec; set fsq_d;
if fsd032A in (1,2) then worried=1; /*yes*/
else if fsd032A=3 then woried=0; /*no*/
else worried=.;
if fsd032B in (1,2) then unout=1; /*yes*/
else if fsd032B=3 then runout=0; /*no*/
else runout=.;
if 1 le fsd190 le 2 then foodstamps=1; /*lt 3 months*/
else if 3 le fsd190 le 6 then foodstamps=2; /*3-6 months*/
else if 7 le fsd190 le 12 then foodstamps=3; /*gt 6 months*/
else foodstamps = .;
run;


proc freq data=foodsec;
tables fsd032A*worried fsd032B*runout fsd190*foodstamps /missing;
run;
```

Next, I want to add in my formats and labels. This is done using the following pattern:

```
proc format;
value formatname
1='1-format for value 1'
2='2-format for value 2';
run;
```

Remember, multiple variables can use the same format. Because the first two variables (WORRIED and RUNOUT) have been recoded following the same pattern, you can use the same format. For this reason, I will only create two new formats, as shown here:

```
proc format;
value evnev
1='1-Ever True'
0='0-Never True';
value months
1='1-Less than 3 Months'
2='2-3 to 6 Months'
3='3-Greater than 6 Months';
run;
```

Note 13.1 ▼

This syntax must be inserted ABOVE the previously created DATA step. Creating the DATA step first is most efficient, allowing you to determine how you will recode your variables (and determine you have recoded your variables correctly), and providing necessary information for creating your formats. SAS, however, must read the formats prior to the DATA step to apply them to your data.

After all of the necessary syntax for applying formats has been written, you will run all of the pieces simultaneously.

Finally, I can go back into the DATA step and add in labels and apply formats. As with labels, this is typically done at the end of the DATA step (but may be done any time after these variables are created):

```
format worried runout evnev. foodstamps months.;
run;
```

Note that although I did not physically write the code in this order, my finished program (including labels) follows these steps:

1. PROC FORMAT for creating your formats
2. Recoding of variables within the DATA step
3. Applying formats with the format statement inside a DATA step

```
1  proc format;
   value evnev
   1='1-Ever True'
   0='0-Never True';
   value months
   1='1-Less than 3 Months'
   2='2-3 to 6 Months'
   3='3-Greater than 6 Months';
   run;

2  data foodsec; set fsq_d;
   if fsd032A in (1,2) then worried=1; /*yes*/
   else if fsd032A=3 then woried=0; /*no*/
   else worried=.;
   if fsd032B in (1,2) then unout=1; /*yes*/
   else if fsd032B=3 then runout=0; /*no*/
   else runout=.;
   if 1 le fsd190 le 2 then foodstamps=1; /*lt 3 months*/
   else if 3 le fsd190 le 6 then foodstamps=2; /*3-6 months*/
   else if 7 le fsd190 le 12 then foodstamps=3; /*gt 6 months*/
   else foodstamps = .;
   label worried='Worried that food would run out before there was
   money to get more' runout='Food ran out before there was money to
   get more' foodstamps='Number of months qualified for food stamp
   assistance';
   format worried runout evnev. foodstamps months.;
3  run;
```

Figure 13.4

When I run a PROC FREQ on these variables, I can now see that the formats have been added to the output:

```
proc freq data=foodsec;
tables worried runout foodstamps;
run;
```

The SAS System

The FREQ Procedure

Worried that food would run out before there was money to get more				
worried	Frequency	Percent	Cumulative Frequency	Cumulative Percent
0-Never True	7565	74.04	7565	74.04
1-Ever True	2653	25.96	10218	100.00

Frequency Missing = 130

Food ran out before there was money to get more				
runout	Frequency	Percent	Cumulative Frequency	Cumulative Percent
0-Never True	8063	78.91	8063	78.91
1-Ever True	2155	21.09	10218	100.00

Frequency Missing = 130

Number of months qualified for food stamp assistance				
foodstamps	Frequency	Percent	Cumulative Frequency	Cumulative Percent
1-Less than three months	92	9.88	92	9.88
2-Three to six months	179	19.23	271	29.11
3-Greater than six months	660	70.89	931	100.00

Frequency Missing = 9417

Figure 13.5

Applying Formats Temporarily in the PROC Step

In most cases, you will wish to apply formats as demonstrated earlier in the DATA step. However, sometimes it is only desirable to apply a format temporarily. In these cases, applying formats in the PROC step may be beneficial. Formats applied in the PROC step will appear in the output resulting from that procedure but will not appear in any subsequent output.

To apply formats in the PROC step, you must first have created the format using the format procedure, as previously demonstrated. The same SAS statement used to permanently apply the format in the DATA step is then used within the PROC step as follows:

```
proc freq data=new;
tables sex;
format sex sexf.;
run;
```

Note that the statement still begins with the SAS keyword FORMAT and is followed by my variable name (in this case SEX) and then the format name (in this case SEXF.). As before, the format name, when referenced in the procedure, must be followed by a period.

Applying formats in a PROC step may be useful if you need to temporarily combine variable categories for display without a permanent recode change. In the previous example, the variable FOODSTAMPS has three categories: 1 = less than 3 months, 2 = 3–6 months, and 3 = greater than 6 months. Suppose that you need to produce output examining those who were eligible for FOODSTAMPS in the groups <=6 months and 7+ months. Instead of creating a new variable, I can create a new format and apply it temporarily in the PROC step.

```
proc format;
value monthtwo
1,2='<6 months'
3='7 or greater months';
run;
```

Note that I have formatted both value 1 and value 2 with the same value label "<6 months." Highlight and run this section of the code. I can then apply the format in the PROC step as follows:

```
proc freq data=foodsec;
tables foodstamps;
format foodstamps monthtwo.;
run;
```

Applying the new format in the procedure produces this output:

foodstamps	Frequency	Percent	Cumulative Frequency	Cumulative Percent
<=6 months	271	29.11	271	29.11
7 or greater months	660	70.89	931	100.00

The SAS System

The FREQ Procedure

Number of months qualified for food stamp assistance

Frequency Missing = 9417

Figure 13.6 Note that the first two categories have been combined.

Although this strategy is useful for temporarily combining categories, it should not be used as a permanent recoding strategy! Values combined in this manner may not perform as expected in all analyses.

Formatting Character Variables

The vast majority of variables you encounter in public health will be numeric. We have covered how to apply formats to numeric variables in the earlier sections. Although we will not cover character formats again in this book, understanding the differences between character and numeric formats may be useful to you someday.

The steps in formatting character variables remain the same as formatting numeric variables but with the following changes:

1. The format name is always proceeded by a $. The dollar sign tells SAS this is a character format.
2. Character variables have characters instead of numbers for the values, and these values should always be listed in quotations.

An example of how this is to be done is shown next:

```
proc format;
value $formatname
'Y'='YES'
'N'='NO';
run;

data new; set old;
format varname $formatname.;
run;
```

Here the underlying values are Y and N (as opposed to 1, 2, 3, etc., that we've used previously). Note that they are listed in quotations. Note also that the format name in both the PROC FORMAT and the DATA step is proceeded by a $.

Suppressing Format Errors

On occasion, your dataset may become separated from its format library, or you may, for some reason, desire not to run your formats during that SAS session. As discussed previously, since we have applied these formats permanently in the DATA step, SAS expects the libraries to be present. When SAS cannot find the permanent or temporary format libraries, you will get an error, and your DATA step will fail to execute. This is obviously a problem. To suppress this error, use the following option.

```
options nofmterr;
```

Let me show you how this works. Let's assume that I have just reopened my SAS program from the previous examples. Since I created a temporary library, I will have to re-run the formats I created to use them. If I fail to do this, I will see the following error in my log:

```
22    data foodsec; set nhanes.fsq_d;
23
24    if fsd032A in (1,2) then worried = 1; *yes;
25    else if fsd032A = 3 then worried = 0; *no;
26    else worried=.;
27
28    if fsd032B in (1,2) then runout = 1; *yes;
29    else if fsd032B = 3 then runout =0; *no;
30    else runout = .;
31
32    if 1 le fsd190 le 2 then foodstamps = 1; *lt 3 months;
33    else if 3 le fsd190 le 6 then foodstamps = 2; *3-6 months;
34    else if 7 le fsd190 le 12 then foodstamps = 3; *gt 6 months;
35
36    label worried = 'Worried that food would run out before there was money to get more'
37    runout = 'Food ran out before there was money to get more'
38    foodstamps = 'Number of months qualified for food stamp assistance';
39
40    format worried runout evnev. foodstamps months.;
                               -------
                                  48
ERROR 48-59: The format EVNEV was not found or could not be loaded.

40 ! format worried runout evnev. foodstamps months.;
                                              -------
                                                 48
ERROR 48-59: The format MONTHS was not found or could not be loaded.

41    run;

NOTE: The SAS System stopped processing this step because of errors.
WARNING: The data set WORK.FOODSEC may be incomplete.  When this step was stopped there were )
         observations and 43 variables.
WARNING: Data set WORK.FOODSEC was not replaced because this step was stopped.
NOTE: DATA statement used (Total process time):
      real time           0.01 seconds
      cpu time            0.01 seconds
```

Figure 13.7

The red text states that the formats I have referenced could not be found. In the green text, I can see that my DATA step has failed to execute.

If I have the temporary format library I created, the best option is just to re-run the formats and try executing the DATA step again. However, if I don't, I can suppress this error by adding the OPTIONS=NOFMTERR (no format error) to my program.

This option works anywhere in your SAS program. I'll place it just above the DATA step in the following example:

```
options nofmterr;

data foodsec; set fsq_d;
if fsd032A in (1,2) then worried=1; /*yes*/
else if fsd032A=3 then woried=0; /*no*/
else worried=.;
if fsd032B in (1,2) then unout=1; /*yes*/
else if fsd032B=3 then runout=0; /*no*/
else runout=.;
if 1 le fsd190 le 2 then foodstamps=1; /*lt 3 months*/
else if 3 le fsd190 le 6 then foodstamps=2; /*3-6 months*/
else if 7 le fsd190 le 12 then foodstamps=3; /*gt 6 months*/
label worried='Worried that food would run out before there was
money to get more' runout='Food ran out before there was money to
get more' foodstamps='Number of months qualified for food stamp
assistance';
format worried runout evnev. foodstamps months.;
run;
```

Although SAS is still identifying the error in the log, as shown in the following, the DATA step executes, and I can continue using it:

```
42    options nofmterr;
43    data foodsec; set nhanes.fsq_d;
44
45    if fsd032A in (1,2) then worried = 1; *yes;
46    else if fsd032A = 3 then worried = 0; *no;
47    else worried=.;
48
49    if fsd032B in (1,2) then runout = 1; *yes;
50    else if fsd032B = 3 then runout =0; *no;
51    else runout = .;
52
53    if 1 le fsd190 le 2 then foodstamps = 1; *lt 3 months;
54    else if 3 le fsd190 le 6 then foodstamps = 2; *3-6 months;
55    else if 7 le fsd190 le 12 then foodstamps = 3; *gt 6 months;
56
57    label worried = 'Worried that food would run out before there was money to get more'
58    runout = 'Food ran out before there was money to get more'
59    foodstamps = 'Number of months qualified for food stamp assistance';
60
61    format worried runout evnev. foodstamps months.;
                            ------
                            484
NOTE 484-185: Format EVNEV was not found or could not be loaded.

61 ! format worried runout evnev. foodstamps months.;
                                            -------
                                            484
NOTE 484-185: Format MONTHS was not found or could not be loaded.

62    run;

NOTE: There were 10348 observations read from the data set NHANES.FSQ_D.
NOTE: The data set WORK.FOODSEC has 10348 observations and 43 variables.
NOTE: DATA statement used (Total process time):
      real time          0.03 seconds
      cpu time           0.04 seconds
```

Figure 13.8

It's important to remember that this will rescue you in a pinch, but all of your output will be printed without the formats applied. For example:

The SAS System

The FREQ Procedure

Number of months qualified for food stamp assistance				
foodstamps	Frequency	Percent	Cumulative Frequency	Cumulative Percent
1	92	9.88	92	9.88
2	179	19.23	271	29.11
3	660	70.89	931	100.00

Frequency Missing = 9417

Figure 13.9

Removing Formats

If you wish to permanently remove previously applied formats from your dataset, you can use one of two methods: (1) removing the format without replacing it or (2) replacing the format.

Removing Formats without Replacement

If you don't want a new format and just want the old one gone, you can use an "empty" format statement as follows:

```
data new; set old;
format sex;
run;
```

Note that the format statement has the keyword format and the variable name sex, but no format is being applied to that variable. This will remove the formats from the specified variables.

If you want to remove formats from an entire dataset, you can use the following shortcut:

where firstvar represents the first variable in the dataset and lastvar represents the last variable in the dataset in creation order.

```
data new; set old;
format firstvar--lastvar;
run;
```

Using PROC CONTENTS with the POSITION option on the foodsec dataset I used previously indicates that SEQN is my first variable and FOODSTAMPS is the final variable:

```
proc contents data=foodsec position;
run;
```

Resulting in the following output:

	Variables in Creation Order				
#	Variable	Type	Len	Format	Label
1	SEQN	Num	8		Respondent sequence number
2	FSD032A	Num	8		HH Worried run out of food
3	FSD032B	Num	8		HH Food didn't last
4	FSD032C	Num	8		HH Couldn't afford balanced meals
5	FSD032D	Num	8		HH Relied on low-cost food for child

Figure 13.10

39	FSD660	Num	8		WIC SP currently get
40	FSD670	Num	8		WIC SP No. mos received
41	worried	Num	8	EVNEV	Worried that food would run out before there was money to get more
42	runout	Num	8	EVNEV	Food ran out before there was money to get more
43	foodstamps	Num	8	MONTHS.	Number of months qualified for food stamp assistance

Figure 13.11   You can also see from this PROC CONTENTS that the variables WORRIED, RUNOUT, and FOODSTAMPS are the only variables with formats applied to them.

Using the following code to remove the formats and then re-running the PROC CONTENTS, I see that the formats have now been removed.

```
data recode; set foodsec;
format seqn--foodstamps;
run;

proc contents data = recode position;
run;
```

39	FSD660	Num	8	WIC SP currently get
40	FSD670	Num	8	WIC SP No. mos received
41	worried	Num	8	Worried that food would run out before there was money to get more
42	runout	Num	8	Food ran out before there was money to get more
43	foodstamps	Num	8	Number of months qualified for food stamp assistance

Figure 13.12

Replacing Formats

If you want a format, but not the one you previously applied, simply create a new format in PROC FORMAT and permanently apply the new format in the DATA step using the syntax format varname formatname.; This will permanently override the previous format.

Using Formats to Customize Category Order[1]

It is often desirable to customize the way your categories appear in output. Formats can allow you to do so. In the previous lesson, I coded 'number of months qualified for food stamps' in increasing order, resulting in the following output:

Number of months qualified for food stamp assistance				
foodstamps	Frequency	Percent	Cumulative Frequency	Cumulative Percent
1-Less than 3 months	92	9.88	92	9.88
2-3 to 6 months	179	19.23	271	29.11
3-Greater than 6 months	660	70.89	931	100.00

Frequency Missing = 9417

Figure 13.13

If I now want to see in my output the frequencies in a different order, I can use my formats to change the order of appearance in the output without changing the recode syntax. By default, SAS sorts values alphanumerically using the underlying numeric value. If I want to sort by the format, I need to add the option ORDER=FORMATTED to the procedure I am running.

---

1   Our many thanks to Dr. Deborah Rosenberg and Dr. Kristin Rankin for teaching this helpful trick, which has saved us many hours and made our SAS ouput so much more useful.

Sorting Output Alphabetically

Let's assume first that I want to sort this variable alphabetically. To do this, I would first need to remove the numbers from the format (text within quotations):

```
proc format;
value months
1 = 'Less than three months'
2 = 'Three to six months'
3 = 'Greater than six months';
run;
```

The alphabetical order for these categories would be (1) greater than six months, (2) less than three months, and (3) three to six months.

To apply these changes, I need to highlight and re-run the edited PROC FORMAT. Then I add to my procedure statement (in this case PROC FREQ) the SAS keyword ORDER and the option FORMATTED as follows:

```
proc freq data = foodsec order = formatted;
tables foodstamps;
run;
```

From my output, I can now see that the order of the categories has been switched to alphabetical:

Number of months qualified for food stamp assistance				
foodstamps	Frequency	Percent	Cumulative Frequency	Cumulative Percent
Greater than six months	660	70.89	660	70.89
Less than three months	92	9.88	752	80.77
Three to six months	179	19.23	931	100.00

Frequency Missing = 9417

Figure 13.14

Sorting Output Using Spaces

Although alphabetical display may be useful in some instances, it is not necessarily desirable in this case, as the categories are now out of their natural ordering. Further, I have lost the value numbers I originally had in my format. To overcome these, I can force the ordering using a series of spaces. SAS will always sort spaces before numbers or letters. For this example, let's assume I would like to view the output in decreasing order.

First, I return to my PROC FORMAT code and add spaces at the beginning of the format (inside the quotation). Since I want to view the output in decreasing order, I want number 3 to appear first. To make this happen, it will need the greatest number of spaces before the format name. I'll add two spaces before the number 3. I'll then add one space before the number 2, and I will not change number 1.

```
proc format;
value months
1='1-Less than three months'
2= ' 2-Three to six months'
3= '  3-Greater than six months';
run;
```

After making the changes, highlight and run this section of the code.

Now when I run my PROC FREQ with the ORDER=FORMATTED option, I can see that my output is in decreasing order:

The SAS System

The FREQ Procedure

Number of months qualified for food stamp assistance				
foodstamps	Frequency	Percent	Cumulative Frequency	Cumulative Percent
3-Greater than six months	660	70.89	660	70.89
2-Three to six months	179	19.23	839	90.12
1-Less than three months	92	9.88	931	100.00

Frequency Missing = 9417

Figure 13.15

This can be particularly useful when creating contingency tables or using logistic regression. In public health, dichotomous variables are commonly used. A dichotomous variable is one that is two levels, typically coded 1 = Yes; 0 = No. This 1 = Yes; 0 = No coding is important in many statistical procedures, including logistic regression, which we will discuss later.

You may recall from your introduction to epidemiology that a contingency (2 x 2) table typically looks like this:

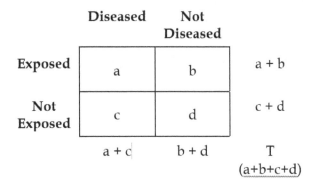

Figure 13.16

To correctly calculate relative risk or odds ratio, we want a table where the "A" cell represents persons who are exposed and also diseased. If, for example, we were interested in looking at the relationship between Medicaid insurance and having a low-birth weight baby, our exposure would be 'having Medicaid' and our outcome "a low-birth weight baby."

Correctly dichotomized variables would be coded as follows:

Medicaid: 1 = Medicaid; 0 = No Medicaid
Low-Birth Weight: 1 = Low-Birth Weight; 0 = Normal Birth Weight

Because SAS sorts alphanumerically, the 0s would be sorted first and our contingency table (or cross tabulation) would look like this:

	Normal Birth Weight	Low Birth Weight	
No Medicaid	a	b	a + b
Medicaid	c	d	c + d
	a + c	b + d	T (a+b+c+d)

Figure 13.17

As such, a relative risk or odds ratio calculated from this data would give the risk of having a normal weight baby if you didn't have Medicaid, as opposed to the desired relationship between having Medicaid and giving birth to a low-birth weight baby. I can reverse this by adding a space in front of the "1" categories, thus forcing SAS to sort them as desired.

Let's work through an example cross tabulation using the variables WORRIED and RUNOUT in the previous section.

Recall that I used this code to format and recode those variables:

```
proc format;
value evnev
1='1-Ever True'
0='0-Never True';
value months
1='1-Less than 3 Months'
2='2-3 to 6 Months'
3='3-Greater than 6 Months';
run;

data foodsec; set fsq_d;
if fsd032A in (1,2) then worried=1; /*yes*/
else if fsd032A=3 then woried=0; /*no*/
```

```
else worried=.;
if fsd032B in (1,2) then unout=1; /*yes*/
else if fsd032B=3 then runout=0; /*no*/
else runout=.;
if 1 le fsd190 le 2 then foodstamps=1; /*lt 3 months*/
else if 3 le fsd190 le 6 then foodstamps=2; /*3-6 months*/
else if 7 le fsd190 le 12 then foodstamps=3; /*gt 6 months*/
label worried='Worried that food would run out before there was mon-
ey to get more' runout='Food ran out before there was money to get
more' foodstamps='Number of months qualified for food stamp assis-
tance';
format worried runout evnev. foodstamps months.;
run;
```

When I run a cross tabulation of the two variables WORRIED and RUNOUT, I get the following table:

The SAS System

The FREQ Procedure

Frequency Percent Row Pct Col Pct	Table of worried by runout		
worried(Worried that food would run out before there was money to get more)	runout(Food ran out before there was money to get more)		
	0-Never True	1-Ever True	Total
0-Never True	7457	108	7565
	72.98	1.06	74.04
	98.57	1.43	
	92.48	5.01	
1-Ever True	606	2047	2653
	5.93	20.03	25.96
	22.84	77.16	
	7.52	94.99	
Total	8063	2155	10218
	78.91	21.09	100.00
Frequency Missing = 130			

Figure 13.18

As I anticipated, the "A" cell is 0, 0 instead of 1, 1. To fix this, I am going to add a space in front of the "1" category in the format procedure.

```
proc format;              ── Add space here
value evnev
1=' 1-Ever True'
0='0-Never True';
run;
```

Highlight and re-run the formats. Next, run a PROC FREQ with the ORDER=FORMATTED option.

```
proc freq data = foodsec order=formatted;
tables worried*runout;
run;
```

Now my output looks correct, and any statistics calculated from the table will use the correct reference groups and predict the desired relationships.

The SAS System

The FREQ Procedure

Frequency Percent Row Pct Col Pct	Table of worried by runout			
	worried(Worried that food would run out before there was money to get more)	runout(Food ran out before there was money to get more)		
		1-Ever True	0-Never True	Total
1-Ever True		2047 20.03 77.16 94.99	606 5.93 22.84 7.52	2653 25.96
0-Never True		108 1.06 1.43 5.01	7457 72.98 98.57 92.48	7565 74.04
Total		2155 21.09	8063 78.91	10218 100.00
Frequency Missing = 130				

Figure 13.19

Summary

Formats are perhaps one of the most effective ways of improving the readability of your output by adding descriptive text to the values for the variables you are working with. Format libraries may be permanent or temporary and each has strengths and weaknesses. For most applications, temporary format libraries are preferred, as they are easily created and transferable from person to person. Like labels, formats may be applied in the DATA step or the PROC step. Applying formats in the DATA step is typically preferred for variables you will be using repeatedly. Formats can be used to customize output display in SAS. This is a useful tool, as it allows you to quickly change the order of variables without the need to recode your data completely.

Lesson #13—Study Questions

1. What is the purpose of formats?
2. How do formats improve SAS output?
3. What are the two steps for creating formats?
4. How are permanent and temporary format libraries different?
5. How does applying formats in the DATA step differ from applying formats in a PROC step?
6. Explain how ORDER=FORMATTED alters the output.
7. Explain how you can use ORDER=FORMATTED to force your preferred sorting of the table.

Lesson 14

Conditional Recoding, Part 1

Learning Outcomes—Students Will:

▶ Identify the difference between simple and conditional recoding,

▶ Demonstrate recoding two raw variables into a single recoded variable, and

▶ Demonstrate using the /LIST option to check recodes from multiple variables.

In Lessons 9, 10, and 11, we learned how to recode categorical and continuous variables. Conditional recoding is an extension of simple recoding where the values of a new variable are dependent on the values of two or more old variables. Conditional recoding is accomplished using the SAS keywords AND & OR. As with simple recoding, conditional recoding is completed in the DATA step and may be combined into a single DATA step with other recoding procedures.

Conditional Recoding with AND

The SAS operator AND is used when two or more conditions must be met. Let's look at an example from the Medical Conditions dataset MCQ_D. Suppose that we are interested in determining the current asthma status of respondents to NHANES. I want to make a new variable that has three categories:

1. Diagnosed with asthma in childhood
2. Diagnosed with asthma in adulthood
3. Never been diagnosed with asthma

In the Medical Conditions dataset, there are two variables that give us this information: MCQ010 and MCQ025. Variable MCQ010 asks, "Has a doctor or other health professional ever told you that you have asthma?" For persons who respond "yes" to this question, Variable MCQ025 asks, "How old were you when you were diagnosed?"

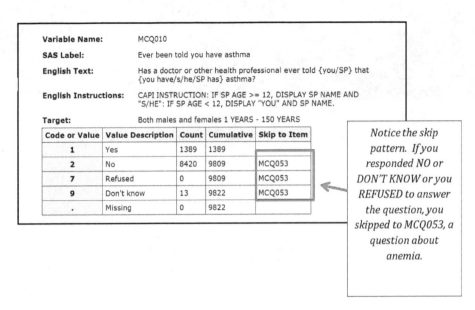

Variable Name: MCQ010

SAS Label: Ever been told you have asthma

English Text: Has a doctor or other health professional ever told {you/SP} that {you have/s/he/SP has} asthma?

English Instructions: CAPI INSTRUCTION: IF SP AGE >= 12, DISPLAY SP NAME AND "S/HE": IF SP AGE < 12, DISPLAY "YOU" AND SP NAME.

Target: Both males and females 1 YEARS - 150 YEARS

Code or Value	Value Description	Count	Cumulative	Skip to Item
1	Yes	1389	1389	
2	No	8420	9809	MCQ053
7	Refused	0	9809	MCQ053
9	Don't know	13	9822	MCQ053
.	Missing	0	9822	

Notice the skip pattern. If you responded NO or DON'T KNOW or you REFUSED to answer the question, you skipped to MCQ053, a question about anemia.

Figure 14.1

Variable Name: MCQ025

SAS Label: Age when first had asthma

English Text: How old {were you/was SP} when {you were/s/he was} first told {he/she} had asthma?

English Instructions: IF LESS THAN 1 YEAR, ENTER 1. CAPI INSTRUCTION: IF SP AGE >= 16, DISPLAY "WERE YOU" AND "YOU WERE". IF SP AGE = 12-15, DISPLAY "WAS {SP}" AND "S/HE WAS". IF SP AGE < 12, DISPLAY "WAS {SP}" AND "YOU WERE". ENTER AGE IN YEARS.

Target: Both males and females 1 YEARS - 150 YEARS

Code or Value	Value Description	Count	Cumulative	Skip to Item
1 to 83	Range of Values	1377	1377	
85	85 years or older	3	1380	
77777	Refused	0	1380	
99999	Don't know	9	1389	
.	Missing	8433	9822	

Figure 14.2

To arrive at our desired variable, I need to combine these two variables into one. As we did in Lesson 8, let's start with a chart. For this example, we will define "childhood" as <18 years of age.

Table 14.1

Value of New Variable	Value(s) of MCQ010	Value(s) of MCQ025
1 = Diagnosed with Asthma in Childhood	1 = Yes	1–17
2 = Diagnosed with Asthma in Adulthood	1 = Yes	18–85
3 = Never Diagnosed with Asthma	2 = No	N/A

As you can see, to fit into category 1 "Diagnosed with Asthma in Childhood," respondents must have said yes, they were diagnosed with asthma AND were 1–17 years of age at the time of diagnosis.

The general format for using the AND statement is as follows:

```
data new; set old;
if varname1 = x and varname2 = x then newvar = x;
else if varname1 = y and varname2 = y then newvar = y;
run;
```

Note that an individual variable can have the same condition in multiple statements, but the complete logical argument (i.e., each SAS statement) if var1 = x and var2 = x, must be unique.

Based on the preceding chart, we use the following code to recode a new variable called asthma conditionally.

```
data one; set nhanes.mcq_d;
if mcq010 = (1) and mcq025 le 17 then asthma = 1; *childhood dx;
else if mcq010 = (1) and 18 le mcq025 le 85 then asthma = 2; *adult dx;
else if mcq010 = 2 then asthma = 3; *no asthma;
run;
```

Note the use of the AND keyword. The first statement says if you have ever been diagnosed with asthma and you were diagnosed when you were less than or equal to 17 (le = less than or equal to), then we classify you as diagnosed with asthma in childhood. You can continue to read the remaining if-then logical statements. Note that MCQ010 = 1 appears in two different statements. This is OK because the second half of the argument is unique in each statement.

Checking Conditional Recoding with the /LIST option

As with simple recoding, it is essential that you carefully check your work. Since we have three variables, a simple cross tabulation that allows you to crosstab only two variables will not work. For this, we will use PROC FREQ with the /LIST option as follows, as we previously did to check calculations in Lesson 10.

```
proc freq data = datasetname;
tables var1*var2*var3 /list missprint;
run;
```

Note the asterisk separating each of the three variables and the option /LIST. The use of the option MISSPRINT in addition to the /LIST option is important. Not including the MISSPRINT option will not allow us to check all possible combinations. Note that you can use this same procedure for more than three variables. Adapting this to the earlier example, my code looks like this:

```
proc freq data= one;
tables mcq010*mcq025*asthma/list missprint;
run;
```

The order of the variables does not technically matter, but for ease, I will add my variables to the tables statement in the order in which they appear in my logic code. Because MCQ025 has a LOT of values (ages ranging from 1–85), I will show you only two sections of the table.

MCQ010	MCQ025	asthma	Frequency	Percent	Cumulative Frequency	Cumulative Percent
1	1	1	361	26.16	361	26.16
1	2	1	79	5.72	440	31.88
1	3	1	65	4.71	505	36.59
1	4	1	49	3.55	554	40.14
1	5	1	81	5.87	635	46.01
1	6	1	58	4.20	693	50.22
1	7	1	52	3.77	745	53.99
1	8	1	57	4.13	802	58.12
1	9	1	37	2.68	839	60.80
1	10	1	66	4.78	905	65.58

Figure 14.3

In this first section of the table, I can see that 361 people who said they had been diagnosed with asthma (MCQ010 = 1) and that they were 1 year of age when diagnosed (MCQ025 = 1) were classified as having a childhood diagnosis of asthma (Asthma = 1).

	11	2	1	0.07	1372	99.42
1	78	2	1	0.07	1373	99.49
1	80	2	1	0.07	1374	99.57
1	82	2	1	0.07	1375	99.64
1	83	2	2	0.14	1377	99.78
1	85	2	3	0.22	1380	100.00
1	99999	.	9	.	.	.
2	.	3	8420	.	.	.
9	.	.	13	.	.	.

Figure 14.4

The second section from the bottom of the table shows us that those who said they had never been diagnosed with asthma (MCQ010 =2) and for whom MCQ025 was missing (because of the skip pattern) were classified as Asthma = 3. This is where the use of missing or missprint is important. In this case, MCQ025 was skipped and therefore missing for those who had never received an asthma diagnosis. If we did not use missprint or missing here, this would be left off of the table, and we would be unable to check the coding of Asthma = 3.

Since the recode appears to have worked as expected, I can now run a simple PROC FREQ to determine the frequency and proportion of persons reporting asthma diagnoses.

The SAS System

The FREQ Procedure

asthma	Frequency	Percent	Cumulative Frequency	Cumulative Percent
1	1079	11.01	1079	11.01
2	301	3.07	1380	14.08
3	8420	85.92	9800	100.00

Frequency Missing = 22

Figure 14.5

Note that I do have 22 people missing for this newly created variable called asthma. Looking back at the output, I can see that these come from the

- 9 respondents who said they had been diagnosed with asthma, but did not know how old they were when diagnosed, and
- 13 respondents who said they did not know if they had been diagnosed with asthma.

1	11	2	1	0.07	1372	99.42
1	78	2	1	0.07	1373	99.49
1	80	2	1	0.07	1374	99.57
1	82	2	1	0.07	1375	99.64
1	83	2	2	0.14	1377	99.78
1	85	2	3	0.22	1380	100.00
1	99999	.	9	.	.	.
2	.	3	8420	.	.	.
9	.	.	13	.	.	.

Figure 14.6

In this case, these appear to be valid missing, although this will not always be the case. Sometimes in conditional recoding, depending on the variable, I can make a logical determination of where "missing" respondents might fit. This will require you to carefully examine all possible combinations for the two (or more) variables you are combining and make educated decisions about the best categorizations. Checking your work is always important, but with conditional recoding, be very conscientious in this process to ensure that every combination of persons is correctly categorized.

Conditional Recoding with OR

The SAS operator OR is used when one of two (or more) conditions must be met. From the Medical Conditions dataset, we can look at an example. Chronic obstructive pulmonary disease (COPD) is a medical condition that includes emphysema and chronic bronchitis. We are interested in identifying the number of NHANES respondents with COPD. From the MCQ_D dataset, there are two variables, MCQ160G "Ever told you had emphysema" and MCQ160K "Ever told you had chronic bronchitis," that can be combined to create a variable for "Ever told you had COPD."

MCQ160G - Ever told you had emphysema

Variable Name:	MCQ160G
SAS Label:	Ever told you had emphysema
English Text:	Has a doctor or other health professional ever told {you/SP} that {you/s/he} . . .had emphysema?
Target:	Both males and females 20 YEARS - 150 YEARS

Code or Value	Value Description	Count	Cumulative	Skip to Item
1	Yes	98	98	MCQ180G
2	No	4872	4970	
7	Refused	0	4970	
9	Don't know	9	4979	
.	Missing	4843	9822	

MCQ160K - Ever told you had chronic bronchitis

Variable Name:	MCQ160K
SAS Label:	Ever told you had chronic bronchitis
English Text:	Has a doctor or other health professional ever told {you/SP} that {you/s/he} . . .had chronic bronchitis?
Target:	Both males and females 20 YEARS - 150 YEARS

Code or Value	Value Description	Count	Cumulative	Skip to Item
1	Yes	290	290	MCQ170K
2	No	4676	4966	
7	Refused	0	4966	
9	Don't know	13	4979	
.	Missing	4843	9822	

Figure 14.7

To create our COPD variable, we will use the following chart:

Table 14.2

COPD	MCQ160G		MCQ160K
1 = Yes	1	OR	1
2 = NO	2	AND	2

Pay close attention to the operators. Note that to recode this, we must use a combination of OR and AND. People have COPD if they EITHER have emphysema OR chronic bronchitis, but they are only free of COPD if they have NEITHER emphysema or chronic bronchitis. Within my DATA step, I can write code to execute the decisions I made in the earlier chart.

```
data copd; set nhanes.mcq_d;
if mcq160G = 1 OR mcq160K = 1 then copd = 1;*yes;
else if mcq160G =2 AND mcq160K = 2 then copd =2; *no;
run;
```

To check our work, I'll run a PROC FREQ using the /LIST MISSPRINT options to ensure that all of the combinations of responses are coded correctly. For ease of conversation, I've used text boxes to label the variables.

```
proc freq data = nhanes.mcq_d;
tables MCQ160G*MCQ160K*COPD/list missprint;
run;
```

Emphysema MCQ160G	Bronchitis MCQ160K	copd	Frequency	Percent	Cumulative Frequency	Cumulative Percent
.	.	.	4843	.	.	.
1	1	1	38	0.77	38	0.77
1	2	1	59	1.19	97	1.95
1	9	1	1	0.02	98	1.97
2	1	1	248	5.00	346	6.97
2	2	2	4613	92.95	4959	99.92
2	9	.	11	.	.	.
9	1	1	4	0.08	4963	100.00
9	2	.	4	.	.	.
9	9	.	1	.	.	.

Frequency Missing = 4859

Figure 14.8

- From this table, we see that there are 4,843 persons who are missing for both variables. Reviewing the codebook, we see that these are a function of skip patterns and question eligibility. This is correct.
- Next, we see that there are 38 people who said, yes, they have both emphysema and bronchitis, and these have been correctly coded as COPD = 1 (Yes).
- There are 59 people who said they have emphysema but not bronchitis; again, these have been correctly coded as COPD = 1 (Yes).

- One person who said he or she had emphysema but didn't know if he or she had bronchitis has been coded as having COPD. Prior to this point, we have largely excluded all refused and don't know. Here is an example of where "don't know" can still be used. To have COPD, they need emphysema OR chronic bronchitis. Since they answered YES to the emphysema question, regardless of whether they have chronic bronchitis or not, they would be classified as having COPD. Here there is no need to exclude this person because they did not know the answer to MCQ160K.
- Let's look at the seventh row in the earlier table. Here we have 11 people who said they did not have emphysema but did not know if they ever had chronic bronchitis. These people have been coded to missing. Is this correct? Let's think about what "don't know" means in this case. Our criteria are that to have COPD you need EITHER emphysema or chronic bronchitis, but to be free of COPD, you need to have NEITHER. For these 11 people, we know they do not have (or did not report having) emphysema, but what can we infer about "don't know" for chronic bronchitis? Without any supplementary data, these people could either truly have COPD or could be free of COPD, but we do not have enough information to confidently assume. To avoid misclassification of these respondents, we will leave them as missing.

Looking through the remainder of the table, we see that our initial recode syntax has worked as we would expect. Always check this table carefully. It is difficult to determine a priori which combinations of responses will appear in the data. Wherever possible, we want to retain as many valid responses as possible.

Finally, I can run a frequency of my newly recoded variable and conclude that 7.1% of the sample has COPD. Notice that I am running these frequencies without MISSING to ensure that my percentages are correctly calculated.

```
proc freq data = copd;
table copd;
run;
```

Have you ever been told you had COPD?				
copd	Frequency	Percent	Cumulative Frequency	Cumulative Percent
1-Yes	350	7.05	350	7.05
2-No	4613	92.95	4963	100.00

Frequency Missing = 4859

Figure 14.9

Summary

Conditional recoding is an important skill in SAS programming, as it allows you to combine multiple variables into a single recoded variable using a combination of AND & OR statements. Because conditional recoding may be somewhat complex, it's important to always map out your recode before you get started and to check your recode using the /LIST option.

Lesson #14—Study Questions

1. What is the difference between simple recoding and conditional recoding?
2. How can you use the operator AND in conditional recoding?
3. How can you use the operator OR in conditional recoding?
4. How do you check conditional recodes?
5. Why is it crucial to use the /LIST MISSPRINT option when cross tabulating your conditional recodes?

Lesson 15

Conditional Recoding, Part 2

Learning Outcomes—Students Will:

▶ Demonstrate combining check-all-that-apply questions into a single recoded variable,

▶ Demonstrate using the /LIST option to plan recoding steps, and

▶ Demonstrate using the /LIST option to check recodes from check-all-that-apply questions.

Conditional Recoding with Check-All-That-Apply Questions

Check-all-that-apply questions are frequently useful in public health but may be tricky to recode. Let's look at the NHANES questionnaire for an example from the Dietary Behavior and Nutrition Module. This is the actual questionnaire text.

First, respondents are asked a question to assess if they consume milk products:

CAPI DISPLAY INSTRUCTIONS: IF SP AGE 7-15 YEARS OLD, DISPLAY "{Next I have some questions about {SP's} eating habits.} First, I'm going to ask about milk products. Do not include their use in cooking. IF SP AGE <= 6 OR => 16 YEARS OLD. DISPLAY "Next I'm going to ask a few questions about milk products. Do not include their use in cooking."

never,	0	(BOX 6)
rarely – less than once a week,	1	
sometimes – once a week or more, but less than once a day, or	2	
often – once a day or more?	3	
VARIED	4	
REFUSED	7	(BOX 6)
DON'T KNOW	9	(BOX 6)

Figure 15.1

Assuming they respond that they rarely, sometimes, or often consume milk products, respondents are then asked what type of milk it usually was. In responding to these questions, participants could choose more than one type of milk, making this a "check-all-that-apply" situation.

DBQ.222   What type of milk was it? Was it usually . . .

IF RESPONDENT CANNOT PROVIDE USUAL TYPE, CODE ALL THAT APPLY

whole or regular,..	10
2% fat or reduced-fat milk..............................	11
1% fat or low-fat milk (includes 0.5% fat milk or	
"low-fat milk" not further specified),.............	12
fat-free, skim or nonfat milk or	13
another type?...	30
REFUSED	77
DON'T KNOW	99

Figure 15.2

In data collection, these sorts of questions usually translate into one variable per response option. In the NHANES dbq data file, there are five variables derived from the responses to this question.

Let's begin by running a frequency procedure on the five variables listed next. We are mostly interested in finding out if there is overlap. That is, did anyone in the sample say he or she usually drank both 2% and whole milk (or any other combination)? We'll use the /LIST and MISSPRINT options covered in the previous section to answer this question.

- DBD222A - You drink whole or regular milk

- DBD222B - You drink 2% fat milk

- DBD222C - You drink 1% fat milk

- DBD222D - You drink fat-free or skim milk

- DBD222U - You drink another type of milk

Figure 15.3

By examining the table that follows, you can see that there are several possible combinations of variables. When there is overlap, a hierarchy must be decided upon. How you decide to categorize these variables is very dependent on the research question. Let's assume that here I am interested in categorizing people by the highest-in-fat milk they usually consume. So if you consume both whole and skim milk, I will consider you a whole milk drinker. I want three categories: 1 = whole milk drinkers, 2 = 1% or 2% milk drinkers, and 3 = skim milk drinkers. Everyone else will be set to missing.

DBD222A	DBD222B	DBD222C	DBD222D	DBD222U	Frequency	Percent	Cumulative Frequency	Cumulative Percent
.	.	.	.	.	1459	14.10	1459	
.	.	.	.	30	310	3.00	1769	
.	.	.	13	.	755	7.30	2524	
.	.	.	13	30	12	0.12	2536	
.	.	12	.	.	657	6.35	3193	
.	.	12	.	30	4	0.04	3197	
.	.	12	13	.	28	0.27	3225	31.17
.	11	.	.	.	3097	29.93	6322	61.09
.	11	.	.	30	31	0.30	6353	61.39
.	11	.	13	.	21	0.20	6374	61.60
.	11	.	13	30	1	0.01	6375	61.61
.	11	12	.	.	65	0.63	6440	62.23
.	11	12	.	30	2	0.02	6442	62.25
.	11	12	13	.	2	0.02	6444	62.27
10	.	.	.	.	3578	34.58	10022	96.85
10	.	.	.	30	40	0.39	10062	97.24
10	.	.	13	.	13	0.13	10075	97.36
10	.	12	.	.	24	0.23	10099	97.59
10	.	12	.	30	1	0.01	10100	97.60
10	11	.	.	.	215	2.08	10315	99.68
10	11	.	.	30	10	0.10	10325	99.78
10	11	.	13	.	1	0.01	10326	99.79
10	11	12	.	.	6	0.06	10332	99.85
10	11	12	.	30	1	0.01	10333	99.86
10	11	12	13	.	1	0.01	10334	99.86
99	.	.	.	.	14	0.14	10348	100.00

Note that you are coded as missing if you don't drink that type of milk.

Figure 15.4

Let's start with a chart.

Table 15.1

New Variable	Old Variable
1 = Whole milk alone or in combination with any other type of milk	DBD222a = 10
2 = 2% or 1% alone or in combination with skim or other milk, excludes whole	DBD222b = 11 OR DBD222C = 12 AND DBD222A ≠ 10
3 = Skim milk only, excludes all reduced or whole fat milk	DBD222d = 13 AND DBD22A ≠10 AND DBD22b ≠ 11 AND DBD222C ≠12
Missing	If DBD222U = 30 AND all others are missing or invalid (refused/don't know)

As you can see, this is somewhat complicated, because there are overlapping conditions. Let's write the code, line by line for each condition. The first line is pretty straightforward. If you drink whole milk alone or with other types of milk, you are a whole milk drinker.

```
data milk; set nhanes.dbq_d;
/*whole milk alone or in combination*/
if dbd222a = 10 then milk = 1;
```

Note that when using a combination of AND & OR in the same SAS statement, you must use parenthetical statements.

For the next line I need a combination of AND & OR statements:

```
/*2% or 1% alone or with skim/other; excludes whole milk*/
else if (dbd222b = 11 OR dbd222c = 12) AND dbd222a ne 10
then milk = 2;
```

For the skim milk line, I need a similar combination of AND & OR statements:

```
/*Skim milk alone*/
else if dbd222d=13 and dbd222a ne 10 and dbd222b ne 11 and dbd222c
ne 13 then milk = 3;
```

Finally, let's make everyone else missing:

```
/*All other missing*/
else if dbd222a in (.,99) and dbd222b = . and dbd222c = . and
dbd222d = . and dbd222u in (.,30)then milk = .;
run;
```

Altogether, this block of syntax looks like this:

```
data milk; set nhanes.dbq_d;
/*whole milk alone or in combination*/
if dbd222a = 10 then milk = 1;
/*2% or 1% alone or with skim/other; excludes whole milk*/
else if (dbd222b = 11 OR dbd222c = 12) AND dbd222a ne 10
then milk = 2;
/*Skim milk alone*/
else if dbd222d=13 and dbd222a ne 10 and dbd222b ne 11 and dbd222c
ne 13 then milk = 3;
```

```
/*All other missing*/
else if dbd222a in (.,99) and dbd222b = . and dbd222c = . and
dbd222d = . and dbd222u in (.,30)then milk = .;
run;
```

Let's check our work using the /LIST MISSPRINT options:

```
proc freq data = milk;
table dbd222a*dbd222b*dbd222c*dbd222d*dbd222u*milk/list missprint;
run;
```

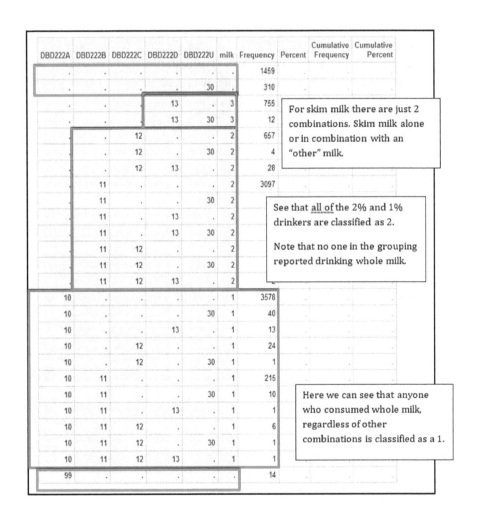

DBD222A	DBD222B	DBD222C	DBD222D	DBD222U	milk	Frequency	Percent	Cumulative Frequency	Cumulative Percent
.	.	.	.	.	.	1459			
.	.	.	.	30	.	310			
.	.	13	.	3		755			
.	.	13	30	3		12			
.	12	.	.	2		657			
.	12	.	30	2		4			
.	12	13	.	2		28			
11	.	.	.	2		3097			
11	.	.	30	2					
11	.	13	.	2					
11	.	13	30	2					
11	12	.	.	2					
11	12	.	30	2					
11	12	13	.	2					
10	.	.	.	1		3578			
10	.	.	30	1		40			
10	.	.	13	.	1	13			
10	.	12	.	.	1	24			
10	.	12	.	30	1	1			
10	11	.	.	.	1	215			
10	11	.	.	30	1	10			
10	11	.	13	.	1	1			
10	11	12	.	.	1	6			
10	11	12	.	30	1	1			
10	11	12	13	.	1	1			
99	.	.	.	.	.	14			

For skim milk there are just 2 combinations. Skim milk alone or in combination with an "other" milk.

See that all of the 2% and 1% drinkers are classified as 2.

Note that no one in the grouping reported drinking whole milk.

Here we can see that anyone who consumed whole milk, regardless of other combinations is classified as a 1.

The two green boxes show the combinations that resulted in missing.

Figure 15.5

Since this appears to be what we wanted, we can add in our formats and labels, run a regular PROC FREQ, and move on.

milk	Frequency	Percent	Cumulative Frequency	Cumulative Percent
		Type of milk used		
1-Whole Milk	3890	45.42	3890	45.42
2-1% or 2% Milk	3908	45.63	7798	91.04
3-Skim Milk	767	8.96	8565	100.00

Frequency Missing = 1783

Figure 15.6

There are some possible ways to shorten the previous code. As you become more familiar with SAS coding, these may become apparent to you. I would always recommend starting out explicitly specifying each condition. Whether you use explicit or reduced code, always be sure you check your work carefully.

Let's take a look at what you might see if you had an error in your code.

Suppose I had written this code:

```
data milk; set nhanes.dbq_d;
/*whole milk alone or in combination*/
if dbd222a = 10 then milk = 1;
/*2% or 1% alone or with skim/other; excludes whole milk*/
else if (dbd222b = 1 OR dbd222c = 12) AND dbd222a ne 10 then
milk = 2;
/*Skim milk alone*/
else if dbd222d=13 and dbd222a ne 10 and dbd222b ne 11 and dbd222c
ne 13 then milk = 3;
/*All other missing*/
else if dbd222a in (.,99) and dbd222b = . and dbd222c = . and
dbd222d = . and dbd222u in (.,30)then milk = .;
*label milk = 'Type of milk used';
*format milk milk.;
run;
```

I use the /LIST MISSPRINT option in PROC FREQ to generate the following table to check my code.

DBD222A	DBD222B	DBD222C	DBD222D	DBD222U	milk	Frequency	Percent	Cumulative Frequency	Cumulative Percent
.	.	.	.	.	.	1459	.	.	.
.	.	.	.	30	.	310	.	.	.
.	.	.	13	.	3	755	.	.	.
.	.	.	13	30	3	12	.	.	.
.	.	12	.	.	2	657	.	.	.
.	.	12	.	30	2	4	.	.	.
.	.	12	13	.	2	28	.	.	.
.	11	.	.	.	.	3097	.	.	.
.	11	.	.	30	.	31	.	.	.
.	11	.	13	.	.	21	.	.	.
.	11	.	13	30	.	1	.	.	.
.	11	12	.	.	2	65	.	.	.
.	11	12	.	30	2	2	.	.	.
.	11	12	13	.	2	2	.	.	.
10	.	.	.	.	1	3578	.	.	.
10	.	.	.	30	1	40	.	.	.
10	.	.	13	.	1	13	.	.	.
10	.	12	.	.	1	24	.	.	.
10	.	12	.	30	1	1	.	.	.
10	11	.	.	.	1	215	.	.	.
10	11	.	.	30	1	10	.	.	.
10	11	.	13	.	1	1	.	.	.
10	11	12	.	.	1	6	.	.	.
10	11	12	.	30	1	1	.	.	.
10	11	12	13	.	1	1	.	.	.
99	.	.	.	.	.	14	.	.	.

Figure 15.7

In looking at the combinations, it appears that all of the milk = 1 and milk = 3 combinations are correctly recoded, but many persons who should be coded as milk = 2 are being coded as missing.

If I carefully examine that line of code, I can find the error. Correcting the error in the code fixes the problem and gives me the original results.

```
/*2% or 1% alone or with skim/other; excludes whole milk*/
else if (dbd222b = 1 or dbd222c = 12) AND dbd222a ne 10 then milk = 2;
```

Whoops! This should be 11 not 1.

Figure 15.8

This error happened to be a pretty easily corrected typo. Errors in logic code may be more difficult to detect in the syntax. **<u>Be sure you carefully check your recodes.</u>** If you identify an unexpected result, revisit your code line by line to determine where the problems arose. Running a line or two of code by itself can sometimes help you more easily isolate errors in coding.

Summary

Check-all-that-apply questions are commonly used but may be very difficult to recode. It is important to first consider how you plan to use the variable and to know the specific questions you would like to answer using the variable. The /LIST MISSPRINT options are very useful for allowing you to understand the degree of overlap of responses in the data and to check your finished work. As with all recoding, it's important to check your work carefully.

Lesson #15—Study Questions

1. What is a check-all-that-apply question?
2. How are check-all-that-apply questions usually denoted in national survey data?
3. Explain the use of the /LIST and MISSPRINT options to examine check-all-that-apply questions.
4. Distinguish between the operators AND & OR for conditional recoding of check-all-that-apply questions.
5. Explain how you check your work to ensure proper recoding.

Lesson 16

DO Groups

Learning Outcomes—Students Will:

▶ Demonstrate conditional recoding through the use of a DO group.

DO Groups

In Lessons 14 and 15, we learned about conditional recoding. In this lesson, you will be introduced to a new operation in SAS called a "DO Group" that can be used to reduce the amount of code necessary for conditional recoding.

In SAS, "DO groups" are used to conditionally process variables. In the previous chapter, we began our exploration of conditional recoding. DO groups are extremely useful conditional recoding tools, as they allow you to process more efficiently, especially complex recodes. DO groups are sometimes also called a DO block.

DO Groups—The Basics

At its simplest, a DO group is an efficient tool for conditional recoding that consists of (1) a conditional DO statement, (2) followed by any number of additional coding steps, and (3) closed with an END statement. The simplest form of the syntax looks something like this:

```
if variable = x then do;
    *additional coding steps;
end;
```

"Variable" can be any valid variable name and "x" any valid value for that variable. Additional logical arguments, including AND, OR, IN ( ), between, etc., can also be used.

DO Groups: A Simple Example

Recall that in Lesson 14, we combined NHANES variables MCQ010 and MCQ025 into a single variable we called ASTHMA that had three levels: 1 = Childhood diagnosis of asthma, 2 = Adult diagnosis of asthma, and 3 = Never diagnosed with Asthma. We used the following SAS code to accomplish this:

```
data one; set nhanes.mcq_d;
if mcq010 = 1 and 1 le mcq025 le 17 then asthma = 1;*childhood dx;
else if mcq010 = 1 and 18 le mcq025 le 85 then asthma = 2;*adult dx;
else if mcq010 = 2 then asthma = 3; *no asthma;
run;
```

Although not a particularly complex recode, this illustrates a good example of the utility of DO groups. Notice that for the first and second lines in the recode syntax, the first condition (MCQ010 = 1) is the same. By using a DO group to first process the "if MCQ010 = 1" condition, we can reduce the amount of text required for the recode.

DO groups begin with the SAS keyword DO and are closed by the SAS keyword END. Between the DO and the END, everything is applied only to respondents who meet the condition(s) set forth in the DO statement. For example, the following code would create the same asthma variable, but with less text:

```
data two; set nhanes.mcq_d;
if mcq010 = 1 then do;

    if 1 le mcq025 le 17 then asthma = 1;
    else if 18 le mcq025 le 85 then asthma = 2;
end;
else if mcq010 = 2 then asthma =3;
run;
```

Note that we begin with the conditional DO statement. If MCQ010 (Have you ever been diagnosed with asthma) = 1 (yes) then do x.

What follows in the green box is all conditioned on MCQ010 being equal to 1, so we only need to complete the logical argument from the original conditional recoding example.

```
if 1 le mcq025 le 17 then asthma = 1;
else if 18 le mcq025 le 85 then asthma = 2;
```

The DO group is closed using SAS keyword END. The entire DO group is repeated next:

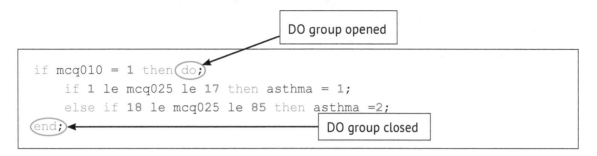

We can check our work using the same procedure as we did in Lessons 13 and 14: PROC FREQ with the /LIST MISSPRINT option. After confirming the code works as we expect, we can format and label our variables and use PROC FREQ to run a frequency on our new variable "asthma." If you compare this to the frequency in Lesson 13, you'll see that both procedures produce exactly the same result.

Ever diagnosed with asthma?				
asthma	Frequency	Percent	Cumulative Frequency	Cumulative Percent
1-Childhood dx	1079	11.01	1079	11.01
2-Adult dx	301	3.07	1380	14.08
3-Never dx with asthma	8420	85.92	9800	100.00

Frequency Missing = 22

Figure 16.1

DO Groups: A More Complex Example

Let's look at an example of a complex conditional recode. In this case, I want to create a recode that categorizes BMI for children. For adults, BMI categorization as underweight, normal weight, overweight, or obese is based only on the value of the BMI. However, because children are still actively growing, and normal adiposity in childhood varies with age and gender, BMI categorizations for children are based on a combination of factors: gender, age, and value of the BMI, which for children is calculated based on percentile distribution on a growth chart.

I have three variables that I wish to combine into one:

- Gender: coded 1 = male; 2 = female
- Age: continuously coded in years
- BMI value

I next need to know what the cutoff values are for the categorizations I wish to achieve. Because this depends on age and gender, I will show you only one example for male children aged seven years. For boys seven years of age, the following BMI categories are defined:

Table 16.1

Category	BMI Value
Underweight	<13.75
Normal Weight	13.75–17.39
Overweight	17.40–19.19
Obese	>=19.20

To recode this in SAS using the techniques covered in Lesson 13, I would write code that looks something like this:

```
data bmi; set one;
if gender = 1 and age = 7 and 0 le BMI lt 13.75 then BMI_Cat = 1;
*Underweight;
else if gender = 1 and age = 7 and 13.75 le BMI lt 17.40 then BMI_
Cat =2; *Normal Weight;
else if gender = 1 and age = 7 and 17.40 le BMI lt 19.20 then BMI_
Cat = 3; *Overweight;
else if gender = 1 and age = 7 and BMI ge 19.20 then BMI_Cat = 4;
*Obese;.
.
.
;
run;
```

Note: I am combining mathematical operators here by using le and lt in my between coding. This can be a useful alteration to the traditional between coding.

Within the context of the DATA step, I have created one line of code for each of the four categorizations. To recode this variable for gender/age/BMI combinations, I would need four lines of code for each combination, requiring an additional 100+ lines of code. A DO group can help us accomplish the same thing with fewer lines of syntax.

A DO group is always created WITHIN the DATA step, so I'll start there, followed by the first conditional DO statement.

```
Data bmi; set one;
if gender = 1 then do;
```

My first DO statement tells SAS that the next block of code should only be applied to male respondents. Next, I want to set the condition that age = 7. I can do this using a second DO group. Note that any number of DO groups can be nested within other DO groups. So I'll add a second conditional DO statement.

```
Data bmi; set one;
if gender = 1 then do;
    if age = 7 then do;
```

Now I can finish the logical argument by adding in the BMI values and closing my DO loop.

```
Data bmi; set one;
if gender = 1 then do;
  if age = 7 then do;
    if 0 le BMI lt 13.75 then BMI_Cat = 1;
    else if 13.75 le BMI lt 17.40 then BMI_Cat =2;
    else if 17.40 le BMI lt 19.20 then BMI_Cat = 3;
    else if BMI ge 19.20 then BMI_Cat = 4;
  end;
```

If you compare the amount of code, you can see that this is more streamlined and efficient but accomplishes the same task.

Note that each DO must have a matching END. In the preceding example, the single END only closes the most recent DO, in this case, if age = 7 then do;. This allows me to continue recoding BMI for male respondents while moving on to different age groups. For example:

```
Data bmi; set one;
if gender = 1 then do;
  if age = 7 then do;
    if 0 le BMI lt 13.75 then BMI_Cat = 1;
    else if 13.75 le BMI lt 17.40 then BMI_Cat =2;
    else if 17.40 le BMI lt 19.20 then BMI_Cat = 3;
    else if BMI ge 19.20 then BMI_Cat = 4;
  end;
  else if age = 8 then do;
    if 0 le BMI lt 13.80 then BMI_Cat = 1;
else if 13.80 le BMI lt 17.90 then BMI_Cat =2;
    else if 17.90 le BMI lt 20.00 then BMI_Cat = 3;
    else if BMI ge 20.00 then BMI_Cat = 4;
end; end;
```

In the second DO block, we are still operating under the first conditional DO (if gender = 1) but have now moved on to age = 8. Again, you should be able to see that this is a more efficient means of carrying out conditional recoding. ***Note*** that the earlier indentations are an organizational strategy but are not required by SAS. Keep in mind when using nested DO groups that an END closes the most recent DO. When you complete your recoding, all DO groups must be closed with an END.

```
                    –
     if gender = 1 then do;                    Every DO, must
       if age = 7 then do;                     have an END
         if 0 le BMI lt 13.75 then BMI_Cat = 1;
         else if 13.75 le BMI lt 17.40 then BMI_Cat =2;
         else if 17.40 le BMI lt 19.20 then BMI_Cat =    3;
         else if BMI ge 19.20 then BMI_Cat = 4;
       end;
       else if age = 8 then do;
         if 0 le BMI lt 13.80 then BMI_Cat = 1;
         else if 13.80 le BMI lt 17.90 then BMI_Cat =2;
         else if 17.90 le BMI lt 20.00 then BMI_Cat = 3;
         else if BMI ge 20.00 then BMI_Cat = 4;
       end; end;                    An END always
                                     closes the most
                                     recent DO
```

Figure 16.2

I would use the same /LIST MISSPRINT option to check this recode as I have done previously.

Summary

DO groups are a useful tool for streamlining conditional recoding. DO groups can produce the same results with fewer lines of code and with syntax that is simpler to follow. DO groups are always opened by the SAS keyword DO and closed by the SAS keyword END. Multiple DO groups can be nested for more complicated recodes.

Lesson #16—Study Questions

1. Explain in a sentence what a DO group is.
2. What are the three elements of a DO group?
3. Give examples of conditional recoding and explain why a DO group is beneficial for conditional recoding.
4. When nesting DO groups, explain the order of the END statements.

Lesson 17

Arrays for Recoding

Learning Outcomes—Students Will:

▸ Explain the difference between a DO group and a DO loop,

▸ Use an array to recode variables, and

▸ Demonstrate using a MACRO variable to reduce syntax.

DO Loops and Arrays

DO loops are similar to DO groups, except they work in an iterative fashion, repeating in a loop for a defined number of repetitions. Although DO loops have other purposes, such as data generation, the primary use for DO loops in data management is in arrays. An array is SAS terminology for a piece of repeating code. DO loops are the engine behind arrays.

At its simplest, a DO loop uses the following syntax:

```
do indexvar = start to stop;
/*whatever you want to DO*/
end;
```

The indexvar is usually a letter, such as "i," and start and stop are replaced by numbers. For example, if I want the loop to repeat six times, I would say:

```
do i = 1 to 6;
/*Whatever you want to DO*/
end;
```

One of the simplest ways to use a DO loop is using an ARRAY. One of the simplest arrays is used when you wish to recode several variables in the same way. For example, if I have several variables that are yes/no variables, I can use an ARRAY and a DO loop to recode them all at once.

In the 2005–2006 NHANES Diet, Behavior, and Nutrition dataset, there are several variables that are yes/no variables. We'll look at variables dbq010, dbq301, dbq330, dbq360, dbq370, dbq400, dbq720, dbq730, and dbq740. All have this structure:

DBQ010 - Ever breastfed or fed breastmilk

Variable Name:	DBQ010
SAS Label:	Ever breastfed or fed breastmilk
English Text:	Now I'm going to ask you some general questions about {SP's} eating habits. Was {SP} ever breastfed or fed breastmilk?
Target:	Both males and females 0 YEARS - 6 YEARS

Code or Value	Value Description	Count	Cumulative	Skip to Item
1	Yes	1484	1484	
2	No	655	2139	DBD040
7	Refused	0	2139	DBD040
9	Don't know	7	2146	DBD040
.	Missing	8202	10348	

Figure 17.1

For all nine variables: 1 = Yes, 2 = No, 7 = Refused, 9 = Don't Know, and . = Missing. We are going to use an array to recode all nine variables into 1 = Yes, 0 = No, Everything else = .

Using simple recoding, I would need to write three lines of code for each of the nine variables I wish to recode. I can simplify the process by creating an array.

```
data new; set nhanes.dbq_d;
if dbq010 = 1 then dbq010R = 1;
else if dbq010 = 2 then dbq010R = 0;
else dbq010R = .;

*etc. for 9 variables*;

run;
```

An array is a list of variables (called elements) that you temporarily group together in SAS, which can then be collectively referred to by the ARRAY name in a DATA step.

SAS Data Management for Public Health

For this task, I will create two ARRAYS in my DATA step:

1. An ARRAY of the nine old variables (dbq010, dbq301, dbq330, dbq360, dbq370, dbq400, dbq720, dbq730, and dbq740).
2. An ARRAY for the nine new variables I want to create. The name of these nine new variables can be any valid variable name, but both ARRAYs should have the variables listed in the same order.

ARRAY syntax is:

```
array arrayname [#elements] list of variables;
```

In this case, I will name the array of old variables YNVARS and the array of new variables YNVARSR. These names are arbitrary, and you can use any name that you like. Since I have nine variables that I want to recode, both ARRAYs will contain nine elements. My ARRAY creation syntax will look like this:

```
array ynvars[9] dbq010 dbq301 dbq330 dbq360 dbq370 dbq400
dbq720 dbq730 dbq740;
array ynvarsR[9] breastmilk govmeals seniorctr kto12
schoollunch schoolbreak  dietguide FGP fiveaday;
```

Now that I've created the ARRAYs, I can use an iterative DO loop to repeat my recoding procedure across the nine elements of the array as follows:

```
do i = 1 to 9;
if ynvars[i] = 1 then ynvarsR[i] = 1;
else if ynvars[i] = 2 then ynvarsR[i] = 0;
else ynvarsR[i]=.;
end;
```

Notice that in the first line of the DO loop, I'm telling SAS that I want to repeat the loop nine times (1 to 9), and that I am using the letter "i" to designate where the repeat should happen. In this DO loop, I am telling SAS that if the old variables (YNVARS[i]) = 1, then the new variables (YNVARSR[i]) = 1. I'm recoding the 2 = No into 0 = No and finally setting everything else to missing. Pay attention to how the ARRAY is referenced in the DO loop: ARRAYname[#elements].

As we did with DO groups, after I have completed the recoding steps I must close the DO loop using an END. When I create the DO loop and say i = 1 to 9, SAS creates a variable called "i" in the database. I will use a DROP statement to eliminate this variable once I have finished the loop (this is not strictly required, but should I choose to use a DO loop again, dropping the "i" allows me to reuse "i" as the indicator).

As compared to the 27 lines of code needed to do this recoding with our simple recoding procedure, we have now reduced our total syntax to 10 SAS statements. My final syntax looks like this:

```
data yn; set nhanes.dbq_d;
array ynvars[9] dbq010 dbq301 dbq330 dbq360 dbq370 dbq400 d
bq720 dbq730 dbq740;
array ynvarsR[9] breastmilk govmeals seniorctr kto12
schoollunch schoolbreak  dietguide FGP fiveaday;
do i = 1 to 9;
if ynvars[i]= 1 then ynvarsR[i] = 1;
else if ynvars[i] = 2 then ynvarsR[i] = 0;
else ynvarsR[i]=.;
end;
drop i;
run;
```

Note that while the number of elements and the index variable must be enclosed in parentheses or brackets, the style of brackets is not important to SAS.

As always, after executing this step, I need to check my recodes. The ARRAY name is not valid outside of the DATA step, so I will need to spell out each of the old and new variable combinations. For demonstration purposes, I am only using two here:

```
proc freq data = yn;
tables dbq010*breastmilk dbq301*govmeals/missprint norow
nocol nopercent;
run;
```

Frequency	Table of DBQ010 by breastmilk				
	DBQ010(Ever breastfed or fed breastmilk)	**breastmilk**			
		.	0	1	Total
	.	8202	0	0	.
	1	0	0	1484	1484
	2	0	655	0	655
	9	7	0	0	0
Total		.	655	1484	2139
Frequency Missing = 8209					

Frequency	Table of DBQ301 by govmeals				
	DBQ301(Community/Government meals delivered)	**govmeals**			
		.	0	1	Total
	.	8778	0	0	.
	1	0	0	56	56
	2	0	1513	0	1513
	7	1	0	0	0
Total		.	1513	56	1569
Frequency Missing = 8779					

Figure 17.2

Checking my recodes, I can see that the variables appear to have been recoded correctly. I can now apply my labels and formats. Although you can use the same format for all nine variables, labels and formats must be individually applied to each variable and may not be applied to the ARRAY name as follows:

```
proc format;
value yn
1 = '1-Yes'
0 = '0-No';
run;
data yn; set nhanes.dbq_d;
array ynvars[9] dbq010 dbq301 dbq330 dbq360 dbq370 dbq400
dbq720 dbq730 dbq740;
array ynvarsR[9] breastmilk govmeals seniorctr kto12 schoollunch
    schoolbreak  dietguide FGP fiveaday;
do i = 1 to 9;
if ynvars[i]=1 then ynvarsR[i] = 1;
else if ynvars[i] = 2 then ynvarsR[i] = 0;
else ynvarsR[i]=.;
end;
drop i;
label breastmilk = 'Ever breastfed or fed breastmilk?'
govmeals = 'Are meals delivered to your home through a goverment
program?'
seniorctr ='Do you eat meals at a community or senior center?'
kto12 = 'Do you attend kindergarten through high school?'
schoollunch = 'Does your school serve lunches?'
schoolbreak = 'Does your school serve breakfast?'
dietguide = 'Have you heard of the dietary guidelines?'
fgp = 'Have you heard of the food guide pyramid?'
fiveaday = 'Have you heard of the five a day program?';
format breastmilk govmeals seniorctr kto12 schoollunch schoolbreak
dietguide fgp fiveaday yn.;
run;
```

As an example, here are the frequencies for two of my recoded variables:

Ever breastfed or fed breastmilk?				
breastmilk	Frequency	Percent	Cumulative Frequency	Cumulative Percent
0-No	655	30.62	655	30.62
1-Yes	1484	69.38	2139	100.00

Frequency Missing = 8209

Are meals delivered to your home through a goverment program?				
govmeals	Frequency	Percent	Cumulative Frequency	Cumulative Percent
0-No	1513	96.43	1513	96.43
1-Yes	56	3.57	1569	100.00

Frequency Missing = 8779

Figure 17.3

More about Arrays

ARRAYs are powerful tools for data management in SAS. In this book, we'll only discuss using an array to recode variables, but you should be aware that the basic ARRAY and DO loop syntax used earlier can be modified to do many different types of tasks, including the following:

- Generating data through a counter (e.g., counting the number of times something happens in a dataset)
- Performing repetitive calculations
- Reading data
- Rotating datasets by changing variables to observations or observations to variables

MACRO Variables

MACRO is an extensive second language in SAS, which consists of imbedded code used to manipulate strings of characters. MACRO programming is a very valuable tool that allows you to quickly and easily invoke a very long, complex line of code. MACROs in most programmer's vernacular refer to very extensive, complicated programs written to accomplish specific tasks, such as complicated calculations and data simulations. These programs are usually written by one person and shared with others. The MACRO, once made available, can be invoked in your dataset using a few words. One example of using a MACRO would be if someone has written code for a complex problem to assess the adequacy of prenatal care from birth certificate data. Many people, in many settings, may be interested in duplicating this analysis using their own data. A MACRO would allow new users to substitute their own variable names without a program and run the index without writing their own, complex code. Although this is very useful, the use of the MACRO language is far beyond the scope of this introductory text. However, MACRO variables are a simple tool that, like DO groups, DO loops, and ARRAYS, can improve coding efficiency.

A MACRO variable is a variable that allows you to substitute one value for another in a SAS program. You can think of a MACRO as an abbreviation. Much like we use the libref as a shortcut for the filepath to our data folder, a MACRO variable can replace almost any command or list in SAS and function as an abbreviation. MACRO variables work anywhere in a SAS program and can make programming very flexible.

At this point in your coding journey, you have probably found yourself writing the same bits of code repeatedly. Anytime you find yourself writing the same line or lines of code repeatedly, this is a good place to use a MACRO variable.

A MACRO variable is created using the %LET statement. Following the %LET, you name the MACRO variable followed by an equal sign and whatever you wish the MACRO to stand for.

For example, when I am conducting my data analysis, I often find that I am referencing the same SAS dataset over and over again through different statistical tests. However, I can reduce the amount of text by creating a MACRO variable that replaces the datasetname. If my dataset name is nhanes. demo_d, I would create a MACRO variable in the following manner:

```
%let name = nhanes.demo_d;
```

After I have highlighted and run this statement, I can invoke the MACRO variable in the dataset by using an ampersand followed by the MACRO variable. In this case "NAME." For example:

```
proc contents data= &name;
run;
proc means data = &name;
var ridageyr;
run;
```

As another example, suppose that I am running multiple procedures on the same list of variables. I can create a MACRO variable that represents this list of variables.

```
%let varlist = indhhinc ridreth1 riagendr;
proc freq data = &name;
tables &varlist;
run;
```

When SAS reads &varlist, it automatically substitutes indhhinc ridreth1 riagendr, and the three frequency tables are produced. Suppose that I wanted to run the same group of procedures on a different set of variables. I can replace the variables included in the %LET statement and rerun the procedure without having to rewrite the entire section of code.

This can be particularly useful if you need to analyze similar datasets over time. Suppose that you are working at the health department, and each year you get a new release of birth certificate data.

Every year, you want to do the exact same analysis on the data. Instead of writing or editing the code each year to complete the analysis, you could use a MACRO variable to represent the year and then only update the %LET statement to execute the analysis.

```
/*Birth Data*/
%let year = 2007;
Proc freq data = dataset;
Where year = &year;
Tables disease;
Run;
%let year1=1999; %let year2=2007;
%let age1 = 10; %let age2 = 19;
proc freq data = birthdata;
where &year1 le birthyr &year2 and &age1 le momsage le &age2;
tables birthyr / out=birthteen;
run;
proc freq data = fetaldeaths;
where &year1 le deathyr le &year2 and &age1 le momsage le &age2;
tables deathyr / out=fetalteen;
run;
```

SAS also has some automatic MACRO variables that you may find useful, particularly, SYSDAY = System Day and SYSDATE = System Date. These MACROs are useful if you need to print out (on paper or electronically) a report. These will print the day and the date the report was run, which may help you later identify the most recent version of output or allow others who are using the data to know that the data is as of a certain date. One way to do this is to use the FOOTNOTE command.

FOOTNOTEs in SAS are similar to titles, but as you may have guessed, they are printed at the bottom of the page. Like titles, footnotes operate anywhere in a SAS program. For example, the following footnotes syntax will display the day and the date the results were executed:

```
Footnote "This report run on &sysday, &sysdate";
```

Gender - Adjudicated				
RIAGENDR	Frequency	Percent	Cumulative Frequency	Cumulative Percent
1	2739	50.06	2739	50.06
2	2732	49.94	5471	100.00

This report run on Monday, 23SEP13

Figure 17.4

MACRO variables are most useful when you have statements or phrases in your code that are repeatedly used. In these situations, using an abbreviated call may improve coding efficiency.

Summary

DO loop, ARRAYs, and MACRO variables are all ways to improve coding efficiency and reduce the number of lines of programming required to accomplish tasks. Although they are sometimes difficult to master, practicing the use of these techniques can pay off very quickly by significantly reducing programming time.

Lesson #17—Study Questions

1. What is an array?
2. What is a DO loop?
3. How is a DO loop different from a DO group?
4. Array syntax includes the number of elements. What are array elements?
5. Explain in general terms when arrays might be useful in coding.
6. What is a macro variable?
7. When are macro variables useful?

Lesson 18

Categorical Data Analysis

Learning Outcomes—Students Will:

> ▶ Explain the hypotheses associated with chi-square tests,
>
> ▶ Explain situations in which chi-square tests may be calculated,
>
> ▶ Demonstrate calculation and interpretation of the chi-square statistic using PROC FREQ, and
>
> ▶ Demonstrate calculation and interpretation of the relative risk and odds ratio using PROC FREQ.

Up to this point, we have learned about a variety of data management techniques in SAS. We've also used several SAS procedures, such as PROC FREQ, PROC MEANS, and PROC UNIVARITE, to get frequencies, percentages, means, medians, and quartile information for our variables. This lesson is going to take these skills a step further by allowing us to explore statistical tests to help us determine the relative importance of our findings. Note that this chapter assumes an understanding of these statistical tests from your introductory statistics and epidemiology classes.

Statistics Using PROC FREQ

The frequency procedure, which is used for categorical variable analysis, can be used to calculate several different types of statistics from cross tabulations. All of these statistics are designed to test the difference between two (or more) groups. We'll cover chi-square, relative risk, and odds ratio.

Chi-Square

The chi-square test is used to examine the association between two categorical variables. The categorical variables may be on any dimension, meaning that they can have any number of levels of categories. The chi-square test measures how far the observed counts in a contingency table are from the counts that would have been expected assuming that there is no relationship between the two variables. If there is no relationship between the two variables, we would expect the distribution of observations to be the same across the levels of the second variable. This "equal distribution" across levels of the second variable is represented by the expected cell count.

$$\text{expected counts} = \frac{\text{row total} \times \text{column total}}{\text{table total}}$$

Once the expected cell counts are calculated, the chi-square test determines how far off the observed counts are from what would be expected, assuming no relationship.

$$\chi^2 = \sum \frac{(\text{observed count} - \text{expected count})^2}{\text{expected count}}$$

The larger the value of chi-square, the more different the observed counts are from what we would expect if there were no relationship. Therefore, a large chi-square indicates that the proportions between the groups are different and that there is some relationship. The question of how large is "large" is dependent on the number of cells a table has; therefore, the chi-square value and its degrees of freedom [(#rows-1)(#columns-1)] are usually compared to a p-value table and a p-value assigned.

The p-value for a chi-square test is interpreted in terms of the null hypothesis. Like all statistics, the null hypothesis for the chi-square test is no association. Formally, this is H0: there is *no* association between the two variables. Conversely, the alternative hypothesis is that some association exists. Formally, HA: there is an association between the two variables. P-values above alpha (typically 0.05) support the null hypothesis, while a p-value less than alpha supports the alternative hypothesis. P-values less than alpha are statistically significant and indicate that random error is an unlikely explanation for the distribution of observations we observed and that there is likely a relationship between the two variables.

In SAS, chi-square is requested as an option (CHISQ) in the tables statement of a PROC FREQ. Note that because chi-square compares the difference of proportions between two variables, a cross tabulation must be used to calculate chi-square.

The syntax that follows is requesting a chi-square test to determine if there is a relationship between marital status and income. My null hypothesis is that there is no relationship between income and marital status, and my alternative hypothesis is that there is an association between marital status and income.

```
Proc freq data = nhanes.demographics;
Tables income*marital/chisq;
Run;
```

This produces my standard cross tabulation output:

Frequency Percent Row Pct Col Pct	Table of income by marital		
	marital(Marital Status)		
income(Household Income)	2-W/D/S/NM	1-Married/Living with a Partner	Total
1-<$15,000	680 10.83 73.83 20.85	241 3.84 26.17 7.99	921 14.67
2-$15,000-$34,999	963 15.34 53.06 29.52	852 13.57 46.94 28.26	1815 28.92
3-$35,000-$64,999	852 13.57 49.71 26.12	862 13.73 50.29 28.59	1714 27.31
4-$65,000 +	767 12.22 41.98 23.51	1060 16.89 58.02 35.16	1827 29.11
Total	3262 51.97	3015 48.03	6277 100.00

Frequency Missing = 4071

Figure 18.1

And gives me a table of statistics:

Statistics for Table of income by marital

Statistic	DF	Value	Prob
Chi-Square	3	253.7612	<.0001
Likelihood Ratio Chi-Square	3	262.3197	<.0001
Mantel-Haenszel Chi-Square	1	215.8433	<.0001
Phi Coefficient		0.2011	
Contingency Coefficient		0.1971	
Cramer's V		0.2011	

Figure 18.2

SAS will give you three different types of chi-square. We'll use the first one. All are roughly the same and 99.9% of cases will give the same qualitative result.

For the relationship between marital status and income, I get a chi-square value of 253.8 for 3 degrees of freedom [(4 – 1)(2 – 1) = 3]. This equates to p<.0001. Because this p-value is <α (typically 0.05), we conclude that if the null hypothesis were true (no relationship), it is unlikely that we would observe such a large difference between observed cell counts and expected cell counts. This supports the alternative hypothesis, and we can conclude that the data suggest that there is a relationship between income and marital status. Chi-square does not give you any information about the nature of the relationship (i.e., which categories are different than we would expect); however, I can determine this by comparing the expected and observed cell counts. Expected cell counts can be requested using the EXPECTED option in the TABLES statement. Note that I have suppressed some of the output using the NOCOL, NOROW, and NOPERCENT options to make the output table easier to read.

```
Proc freq data = nhanes.demographics;
Tables income*marital/chisq expected nocol norow nopercent;
Run;
```

Frequency Expected	Table of income by marital		
		marital(Marital Status)	
income(Household Income)	2-W/D/S/NM	1-Married/Living with a Partner	Total
1-<$15,000	680 478.62	241 442.38	921
2-$15,000-$34,999	963 943.21	852 871.79	1815
3-$35,000-$64,999	852 890.72	862 823.28	1714
4-$65,000 +	767 949.45	1060 877.55	1827
Total	3262	3015	6277
Frequency Missing = 4071			

Figure 18.3

Comparing the observed to the expected counts in the preceding table, I can see that more than expected widowed, divorced, and separated persons had income <$15,000, and more than expected married/cohabitating persons had income >=$65,000. From this we can conclude that those who are married or living with a partner appear to have larger incomes than those who are living singly.

Relative Risk and Odds Ratio

Recall from your introduction to epidemiology course that the relative risk (also called risk ratio or rate ratio) is a ratio of the incidence in the exposed and the incidence in the unexposed.

$$\textit{Relative Risk} = \frac{I_e}{I_u} = \frac{a / (a + b)}{c / (c + d)}$$

The relative risk is calculated from a contingency table and can be used to compare the risk of an outcome in two groups. When calculating the relative risk from PROC FREQ in SAS, both the exposure and the outcome variables should be dichotomous (two levels). A different procedure must be used to calculate a measure of association for variables with more than two levels. The relative risk uses a null and alternative hypothesis, like other statistics. As always, the null hypothesis is that there is no association between the exposure and the outcome. This means that the incidence rate in those with the exposure will be equal to the incidence rate in those without the exposure. The alternative hypothesis would be that those with the exposure have a different incidence rate than those without the exposure.

Suppose that I am interested in knowing if people who drink whole milk are more or less likely to be familiar with the (now defunct) food guide pyramid. My null hypothesis is that those who drink whole milk have the same likelihood of being unfamiliar with the food guide pyramid as those who do not drink whole milk, or that the relative risk will be 1.0. I have recoded the variable to be 1 = drinks whole milk; 0 = does not drink whole milk, and a second variable that is 1 = has not heard of the food guide pyramid; 0 = has heard of the food guide pyramid. I am using "has not heard" as the outcome of interest because my hypothesis is that those who drink full fat milk are less likely to have heard of the food guide pyramid.

Like chi-square, the relative risk is requested in the options list for the cross tabulation. The option is RELRISK.

```
proc freq data = nhanes.diet order = formatted;
tables wholemilk*foodpyr/relrisk norow nocol nopercent;
run;
```

As before, the frequency procedure produces the 2 x 2 table. I have suppressed the percentages to make the table easier to read. Note that the setup of the table is very important in interpreting the relative risk. The outcome of interest and the exposure of interest should intersect in the "A" cell (top left) of the contingency table. If your table is not oriented as you like, you can use the ORDER=FORMATTED option to change the order of the rows and/or columns (see Lesson 13).

Frequency	Table of wholemilk by foodpyr		
		foodpyr(Unfamiliar with food pyramid)	
wholemilk(Drinks whole milk?)	1-Yes	0-No	Total
1-Yes	728	1244	1972
0-No	938	3161	4099
Total	1666	4405	6071
Frequency Missing = 4277			

Figure 18.4

Recall that the relative risk is calculated as [(a/(a + b)) / (c/(c + d))]. So in this case,

$$Relative\ Risk = \frac{\dfrac{728}{1972}}{\dfrac{938}{4099}} = 1.61.$$

As with chi-square, in addition to the contingency table, SAS produces a table of statistics. Included in the statistics are the measure of the strength of the association (risk ratio or odds ratio) and the 95% confidence interval for the estimate.

Statistics for Table of wholemilk by foodpyr

Estimates of the Relative Risk (Row1/Row2)			
Type of Study	Value	95% Confidence Limits	
Case-Control (Odds Ratio)	1.9721	1.7545	2.2168
Cohort (Col1 Risk)	1.6132	1.4884	1.7486
Cohort (Col2 Risk)	0.8180	0.7878	0.8494

Figure 18.5

The second line produces the risk ratio we calculated earlier, which can be interpreted as those who drink whole milk are 1.61 times as likely to be unfamiliar with the food guide pyramid as compared to those who do not drink whole milk. Since the confidence interval for this estimate does not include 1.0, we conclude that this is a statistically significant result, which means that this is unlikely to be attributable to random error.

It is important to note that the relative risk is based on incidence rates. As such, this statistic should only be used in studies that use longitudinal data. Although, I have used data from NHANES (a cross-sectional study) to illustrate the use of the relative risk, an odds ratio is a more appropriate measure of association for NHANES.

The other values in the table of statistics noted earlier may also be useful. The first is the ODDS RATIO. The odds ratio is an estimate of the risk ratio in situations where incidence is not available and is calculated as the cross-product ratio (ad/bc).

$$Odds\ Ratio = \frac{ad}{bc} = \frac{728 \times 3161}{1244 \times 938} = 1.97$$

Since NHANES is a cross-sectional study and therefore by definition does not have incidence data, we should use odds ratio when analyzing this data. The odds ratio has a similar interpretation to the risk ratio. However, rather than calculating the "risk" of some outcome, we are more correctly calculating the "odds" of some outcome. I would interpret this odds ratio as those who drink whole milk had 1.97 times the odds of being unfamiliar with the food guide pyramid as compared to those who do not drink whole milk.

The last row is the relative risk for the opposite relationship. This is interpreted as those who DO NOT drink whole milk are 0.82 times as likely to be familiar with the food guide pyramid as compared to those who do drink whole milk. If you have oriented your contingency table correctly, you will not need this value. The orientation of your contingency table and the use of statistics is dependent on your research question. Although SAS will output many values for you, it cannot tell you which values are appropriate for the research question or your data.

The same statistics can be requested using the option CMH in the TABLES statement. As you can see, the same values are calculated using either CMH or RELRISK.

```
proc freq data = nhanes.diet order = formatted;
tables wholemilk*foodpyr/cmh;
run;
```

Estimates of the Common Relative Risk (Row1/Row2)				
Type of Study	Method	Value	95% Confidence Limits	
Case-Control	Mantel-Haenszel	1.9721	1.7545	2.2168
(Odds Ratio)	Logit	1.9721	1.7545	2.2168
Cohort	Mantel-Haenszel	1.6132	1.4884	1.7486
(Col1 Risk)	Logit	1.6132	1.4884	1.7486
Cohort	Mantel-Haenszel	0.8180	0.7878	0.8494
(Col2 Risk)	Logit	0.8180	0.7878	0.8494

Figure 18.6

Summary

Chi-square, relative risk, and odds ratio are three measures for evaluating the unadjusted relationship between categorical variables. Chi-square can be used for categorical variables with multiple values; however, odds ratio and relative risk can only be calculated from PROC FREQ when both comparison variables have only two levels. The frequency procedure will generate all three of these statistics as options in the TABLES statements using CHISQ, RELRISK, and CMH options. Careful interpretation of these tests will help you to evaluate relationships and rule out the presence of random error as an alternative explanation for the observed associations.

Lesson #18—Study Questions

1. What is a chi-square test?
2. How are expected values calculated?
3. The chi-square test is based on the hypothesis that: _____.
4. What does a chi-square test with a p-value < alpha indicate?
5. What does a chi-square test with a p-value > alpha indicate?

6. What kind of variables can you use the chi-square test for?
7. A chi-square test is requested in what SAS procedure?
8. What is the SAS keyword for requesting a chi-square test?
9. Explain the difference between the relative risk and the odds ratio.
10. What SAS procedure is used for requesting relative risk or odds ratio.
11. What are the two SAS keywords that produce odds ratios and relative risks?
12. How do you interpret a relative risk or odds ratio?
13. What is the interpretation of the confidence interval?

Lesson 19

Stratified and Subpopulation Analysis

Learning Outcomes—Students Will:

> ▶ Explain the difference between subpopulation and stratified analysis,
>
> ▶ Demonstrate the use of the WHERE criterion for subpopulation analysis, and
>
> ▶ Demonstrate the use of the BY criterion for stratified analysis.

There are many situations in which it is desirable to conduct analysis separately for different groups of people or only among a subset of the population. These are easily accomplished in SAS using the SAS keywords BY and WHERE.

Stratified Analysis

Stratified analysis conducts separate data analyses for different segments of the population. For example, perhaps I want to analyze men separately from women or children separately from adults. There are many reasons this is desirable. Recall from your introduction to epidemiology course that stratification is a useful tool for evaluating and controlling for confounding. In other cases, it may simply be informative to separately understand the distribution of exposures or outcomes among people with different characteristics.

In SAS, the BY statement, which can be added to most PROCs, allows you to simultaneously conduct analyses in multiple segments of the population. <u>The BY statement produces one table or model output for each level of the stratification (BY) variable.</u> For example, if I stratified by gender, two tables will be produced (male and female). However, if I stratify by year in school, four tables will be produced (freshman, sophomore, junior, and senior).

One important reminder, when using a BY statement, the data must be SORTED (using PROC SORT) by the BY variable. You will recall that we had to SORT the data previously when using the BY criterion for merging datasets.

I will give examples of stratified analysis using PROC FREQ and PROC MEANS. However, the same applies to other SAS procedures. Recall that sorting the data is done with a PROC SORT:

```
proc sort data = datsetname;
by sort variable;
run;
```

While stratified analysis is completed with PROC FREQ or PROC MEANS:

```
proc freq data = datasetname;
tables variable(s);
by stratification variable;
run;

proc means data = datasetname;
var variable(s);
by stratification variable;
run;
```

Note that in both cases, the PROC code is the same as it has been up to this point with the only addition being the BY statement. The BY statement consists of the SAS keyword BY followed by the name of the variable you wish to stratify by.

Suppose that I wanted to look at the distribution of income. Because income (in this case) is a categorical variable, I will use PROC FREQ. Using the standard PROC FREQ as indicated next, I get a single table telling me the distribution of income in the population.

```
proc freq data = recode;
tables income;
run;
```

Household Income				
income	Frequency	Percent	Cumulative Frequency	Cumulative Percent
1-<$15,000	1486	15.28	1486	15.28
2-$15,000-$34,999	2950	30.34	4436	45.62
3-$35,000-$64,999	2536	26.08	6972	71.70
4-$65,000 +	2752	28.30	9724	100.00

Now suppose that I am interested in seeing if the distribution differs by gender. Adding a BY statement produces one table for each gender.

```
proc sort data = recode;
by gender;
run;
proc freq data = recode;
tables income;
by gender;
run;
```

Gender=1-Male

Household Income

income	Frequency	Percent	Cumulative Frequency	Cumulative Percent
1-<$15,000	537	10.21	537	10.21
2-$15,000-$34,999	1528	29.04	2065	39.25
3-$35,000-$64,999	1437	27.31	3502	66.57
4-$65,000 +	1759	33.43	5261	100.00

Figure 19.1

Gender=2-Female

Household Income

income	Frequency	Percent	Cumulative Frequency	Cumulative Percent
1-<$15,000	949	21.26	949	21.26
2-$15,000-$34,999	1422	31.86	2371	53.13
3-$35,000-$64,999	1099	24.62	3470	77.75
4-$65,000 +	993	22.25	4463	100.00

Figure 19.2

Note that the tables are identified by the Gender = 1 or Gender = 2 titles located above the tables. Similarly, I can do a stratified analysis for PROC MEANS.

```
proc sort data = recode;
by income;
run;
proc means data = recode;
var ridageyr;
by income;
run;
```

After sorting the data by income (my stratification variable), I can produce one table of means for each of the four income categories.

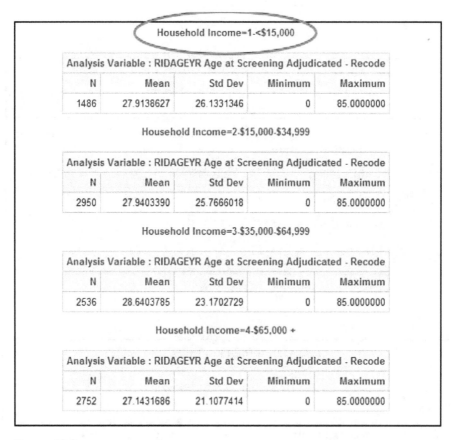

Figure 19.3

Note that if you have missing data for your variable, a table of statistics for the missing level will also be produced. In most cases, this can be ignored.

Subpopulation Analysis

Subpopulation analysis is similar to stratified analysis, except that in this case, we conduct analysis only for a subset of the population. Subpopulation analysis is done by adding WHERE criteria to most PROCs. Unlike the BY statement, use of the WHERE criterion does not require sorting of the data.

Subpopulation analysis for PROC FREQ and PROC MEANS:

```
proc freq data = datasetname;
tables variable(s);
where subpopvariable = x;
run;

proc means data = datasetname;
var variable(s);
where subpopvariable = x;
run;
```

The variables in the tables/var and where statements may be the same or different. If I return to my example of looking at income and gender, I can easily produce a table of income distribution for men only. Unlike the BY statement, the table is not inherently labeled with the subpopulation. Looking at the cumulative frequency will help you determine if the analysis was done correctly. It may also be wise to use titles for subpopulation analysis so that each output table is clearly identified.

```
proc freq data = recode;
tables income;
where gender =1;  *income for males only;
run;
```

Household Income				
income	Frequency	Percent	Cumulative Frequency	Cumulative Percent
1-<$15,000	371	9.49	371	9.49
2-$15,000-$34,999	1111	28.42	1482	37.91
3-$35,000-$64,999	1049	26.84	2531	64.75
4-$65,000 +	1378	35.25	3909	100.00

Figure 19.4

Household Income				
income	Frequency	Percent	Cumulative Frequency	Cumulative Percent
1-<$15,000	537	10.21	537	10.21
2-$15,000-$34,999	1528	29.04	2065	39.25
3-$35,000-$64,999	1437	27.31	3502	66.57
4-$65,000 +	1759	33.43	5261	100.00

Figure 19.5

Note that this is the same table created for men using the BY statement. The only difference is that this time, no female table was created. One nice feature of the WHERE criterion is that multiple criteria may be specified by using AND & OR statements. For example, this code requests income for men who are <45 years of age.

```
proc freq data = recode;
tables income;
where gender = 1 and ridageyr le 45;
run;
```

We can again see an example of subpopulation analysis using PROC MEANS. Here I am requesting the mean age of those who have income ≤$34,999:

```
proc means data = recode;
var ridageyr;
where income in (1,2);
run;
```

Analysis Variable : RIDAGEYR Age at Screening Adjudicated - Recode				
N	Mean	Std Dev	Minimum	Maximum
4436	27.9314698	25.8870200	0	85.0000000

Figure 19.6

As before, a single table is produced. You can see that the n = 4,436 observations used in this subpopulation analysis is equal to the 1,486 persons in the income category #1 plus the 2,950 persons in the income category #2.

Summary

Both subpopulation and stratified analysis are very useful for focusing your analysis on specific subsets of the population. This technique is particularly useful in the context of a broader analysis. Although the overall project may be focused on (for example) US citizens, understanding how the distribution differs among specific demographic groups is useful and important. In addition, stratified analysis may help you to evaluate the potential for confounding bias in your results.

Lesson #19—Study Questions

1. What is the difference between stratified analysis and subpopulation analysis?
2. What SAS statement is used for requesting a stratified analysis?
3. What SAS statement is used for requesting a subpopulation analysis?

Lesson 20

Logistic Regression

Learning Outcomes—Students Will:

▶ Demonstrate the use of PROC LOGISTIC to calculate crude and adjusted odds ratios, and

▶ Demonstrate the calculation and interpretation of the relative risk and odds ratio using PROC FREQ.

Logistic Regression

Logistic regression is an extremely useful analysis tool in public health. Like the linear regression you learned about in statistics, logistic regression "regresses" one or more determinants (independent variables/risk factors) against a single outcome (dependent variable). Linear regression uses a continuous outcome variable making its utility limited in public health where dichotomous or categorical outcomes are much more common. In public health, we are often interested in dichotomous outcome variables (dead vs. alive, diseased vs. not diseased, pregnant vs. not pregnant). Using logistic regression, we can predict the odds of the dichotomous outcome from a set of several predictor variables.

Briefly, logistic regression is a regression method that uses the logit transformation to force the prediction equation to predict values between 0 and 1. Logistic regression predicts the natural log of the odds of being a subject in the group of interest as compared to the reference group. The resulting coefficients can be exponentiated to estimate the odds ratio. Unlike the odds ratio we calculated from PROC FREQ, PROC LOGISTIC can give you the odds ratios for multiple exposure groups (i.e., more than two). However, the outcome variable must always be dichotomous:

- The dependent or outcome variable in logistic regression must be binary.
- The independent or exposure variables may be binary, categorical or continuous.

PROC LOGISTIC syntax for a simple crude logistic regression is as follows:

```
proc logistic data = datasetname;
class riskfactor(s);
model outcome = riskfactor(s);
run;
```

The first line looks like all of the PROCs we have used to this point. The ORDER=FORMATTED option (Lesson 13) is available in this procedure and may be useful to ensure the correct odds ratios are calculated.

CLASS is a keyword in SAS that specifies which of the predictor variables (those on the right side of the model statement) are categorical. SAS will not correctly calculate odds ratios for categorical variables that are not present in the CLASS statement.

The MODEL statement is where you tell SAS the equation you would like it to regress. The model statements should always be constructed as the keyword MODEL, followed by the outcome variable and equal sign, and then a list of all of the risk factors or predictor variables for your model (e.g., MODEL OUTCOME = RISKFACTOR(s)).

Let's look at an example. I want to predict the odds of having a stroke for people who smoke. I have created two variables: stroke is my outcome and smoking status is the exposure. It is always a good idea to look at the frequency or means tables for your data before you proceed with regression analysis or other statistical tests. Looking at the frequencies, I can see that approximately 4% of my population had experienced a stroke and about 19% were current smokers.

Ever Had a Stroke				
stroke	Frequency	Percent	Cumulative Frequency	Cumulative Percent
0-Did not have a Stroke	5505	96.34	5505	96.34
1-Had a Stroke	209	3.66	5714	100.00
Frequency Missing = 5				

Smoking Status				
smoke	Frequency	Percent	Cumulative Frequency	Cumulative Percent
1-Current Smoker	1068	18.71	1068	18.71
2-Former Smoker	1319	23.11	2387	41.82
3-Never Smoker	3321	58.18	5708	100.00
Frequency Missing = 11				

Figure 20.1

Since logistic regression will produce the odds ratios, my null hypotheses for the logistic regression model are the same as for other odds ratios. Here my null hypothesis is that there is no association between smoking status and having a stroke. My alternative hypothesis is that there is a relationship between smoking status and having a stroke.

To look at the association between these two variables using logistic regression, I would use the following syntax.

```
proc logistic data = cvd;
class smoke/param=ref;
model stroke = smoke;
run;
```

Note that I've included SMOKE in the CLASS statement because it is a categorical variable. I am also using the option PARAM=REF in the class statement. This option tells SAS to use reference parameterization, rather than effect parameterization. The merits of reference parameterization are beyond the scope of this class, but I generally recommend using reference parameterization.

Let's take a look at the output. The first step is to understand which odds SAS is predicting. I can do this by looking at the Response Profile in the output.

Response Profile		
Ordered Value	stroke	Total Frequency
1	0-Did not have a Stroke	5495
2	1-Had a Stroke	209

Probability modeled is stroke='0-Did not have a Stroke'.

Figure 20.2

By default, SAS will predict the odds of the outcome level that has the lowest value. In this example, since my variable is coded '0 = did not have a stroke' and '1 = had a stroke,' SAS is predicting the odds of NOT having a stroke. I probably want to reverse this to predict the odds of HAVING a stroke, so I need to use the DESCENDING option or ORDER=FORMATTED (Lesson 13).

```
proc logistic data = cvd descending;
class smoke/param=ref;
model stroke = smoke;
run;
```

By adding the descending option, I can see that SAS is now predicting the odds of HAVING a stroke. Confirming that your response profile is modeling the probability of the outcome of interest is essential or all of the remaining output will be wrong.

Response Profile		
Ordered Value	stroke	Total Frequency
1	1-Had a Stroke	209
2	0-Did not have a Stroke	5495

Probability modeled is stroke='1-Had a Stroke'.

Figure 20.3

The choice to use DESCENDING or ORDER=FORMATTED largely depends on if you need to change the order of your predictor variables as well as the outcome, or the outcome only. DESCENDING only reverses the order of the OUTCOME variable. ORDER=FORMATTED allows you to select the order of OUTCOME and EXPOSURE variables.

Next, we need to look at the Class Level Information table. The class level information table tells you which level of the variable is being used as the reference group. By default, SAS chooses the highest value. Since my smoking variable is coded 1 = Current Smoker, 2 = Former Smoker, and 3 = Never Smoker, SAS will use never smoker as the reference group. Since this is likely the group at the lowest risk of a stroke, in this case, we can proceed.

Class Level Information			
Class	Value	Design Variables	
smoke	1-Current Smoker	1	0
	2-Former Smoker	0	1
	3-Never Smoker	0	0

Figure 20.4   The reference group will be denoted by all 0s or all −1s.

However, if I had coded my variable as 0 = Never, 1 = Current, and 2 = Former, then former would be used as the reference group, and I would probably want to use ORDER = FORMATTED to choose my reference categories.

Type 3 Analysis of Effects

Effect	DF	Wald Chi-Square	Pr > ChiSq
smoke	2	28.8217	<.0001

Analysis of Maximum Likelihood Estimates

Parameter		DF	Estimate	Standard Error	Wald Chi-Square	Pr > ChiSq
Intercept		1	-3.6513	0.1105	1091.5394	<.0001
smoke	1-Current Smoker	1	0.6585	0.1811	13.2178	0.0003
smoke	2-Former Smoker	1	0.8309	0.1629	26.0192	<.0001

Figure 20.5

The Type 3 Analysis of Effects tables test the null hypothesis that the exposure is not associated with the outcome. Specifically, here, that smoking status is not associated with odds of a stroke. Because this p-value is less than alpha (typically 0.05), this supports the alternative hypothesis that there is an association between smoking status and having a stroke. This p-value, however, doesn't tell us which levels of smoking are associated with an increased (or decreased) odds of stroke.

The Analysis of Maximum Likelihood Estimates table give the coefficient estimates. When using reference parameterization, the estimates are the natural log of the odds ratio. Thus exponentiating these values will yield the odds ratio. Helpfully, SAS takes this step and calculates these for you in the next table.

The last table in the output that is interesting to me is the Odds Ratio Estimates table.

Odds Ratio Estimates

Effect	Point Estimate	95% Wald Confidence Limits	
smoke 1-Current Smoker vs 3-Never Smoker	1.932	1.355	2.755
smoke 2-Former Smoker vs 3-Never Smoker	2.295	1.668	3.159

Figure 20.6

The odds ratio estimates table gives you the odds of having a stroke for current versus never smokers (1.932) and the odds of having a stroke for former versus never smokers (2.295), and the corresponding confidence intervals. If I were to calculate these by hand from the 3 x 2 contingency table, I would get the same answers.

Adjusted Logistic Regression

One of the major advantages of logistic regression over the standard calculation of the odds ratio is that logistic regression allows you to calculate the odds ratios for your exposure-disease relationship while adjusting for a variety of other factors. Although the aforementioned results indicate that smoking status is associated with having a stroke, there are many other factors that may influence the risk of stroke. To help us isolate the effect of smoking status, we may wish to control or adjust for other factors known to increase (or decrease) the risk of stroke.

To adjust a logistic regression model, the following syntax is used:

```
proc logistic data = nhanes.demographics order = formatted;
class riskfactor confounders;
model outcome = riskfactor confounders;
run;
```

Let's first look at the new variables gender and age. You can see that I have six categories for age and two for gender. I have already learned from SAS is that it uses the highest value as the reference category for predictor variables. In this case, SAS will default to 65+ years on age and female for gender. Since males have a higher risk of stroke than females, this is fine, but age is strongly correlated with stroke risk, so I'd like to use the youngest age group as the reference category. I can do this using ORDER = FORMATTED or the (REF=) option in the CLASS statement.

Gender				
gender	Frequency	Percent	Cumulative Frequency	Cumulative Percent
1-Male	2747	48.03	2747	48.03
2-Female	2972	51.97	5719	100.00

Age in Years				
age	Frequency	Percent	Cumulative Frequency	Cumulative Percent
1-20-24 years	441	7.71	441	7.71
2-25-34 years	1035	18.10	1476	25.81
3-35-44 years	951	16.63	2427	42.44
4-45-54 years	956	16.72	3383	59.15
5-55-64 years	958	16.75	4341	75.90
6-65+ years	1378	24.10	5719	100.00

Figure 20.7

First let me demonstrate the default ordering to confirm my reference groups.

```
proc logistic data = cvd;
class smoke age gender /param=ref;
model stroke = smoke age gender;
run;
```

The preceding code produces this class level information table:

Class	Value	Design Variables				
smoke	1-Current Smoker	1	0			
	2-Former Smoker	0	1			
	3-Never Smoker	0	0			
age	1-20-24 years	1	0	0	0	0
	2-25-34 years	0	1	0	0	0
	3-35-44 years	0	0	1	0	0
	4-45-54 years	0	0	0	1	0
	5-55-64 years	0	0	0	0	1
	6-65+ years	0	0	0	0	0
gender	1-Male	1				
	2-Female	0				

Figure 20.8

The Class Level Information table confirms that the default reference groups are never smokers, 65+ year of age, and female.

Using ORDER= FORMATTED

Recall from Lesson 13 that you can add spaces in front of your format to force the ordering. Next are the formats I have written.

I have highlighted the levels I wish to use as reference groups. Now I will go in and add spaces to the age and stroke variables to force the ordering to appear as I would like. **Remember that the reference group is closest to the quotes.**

```
proc format;
value sex
1 = '1-Male'
2 = '2-Female';
value stroke
1 = '1-Had a Stroke'
0 = '0-Did not have a Stroke';
value smoke
1 ='1-Current smoker'
2 = '2-Former Smoker'
3 = '3-Never Smoker';
value age
1 = '1-20-24 years'
2 = '2-25-34 years'
3 = '3-35-44 years'
4 = '4-45-54 years'
5 = '5-55-64 years'
6 = '6-65+years';
run;
```

Note that I have not altered the gender and smoke variables as the SAS defaults will order these as needed.

```
proc format;
value sex
1 = '1-Male'
2 = '2-Female';
value stroke
1 = ' 1-Had a Stroke'
0 = '0-Did not have a Stroke';
value smoke
1 ='1-Current smoker'
2 = '2-Former Smoker'
3 = '3-Never Smoker';
value age
1 = '1-20-24 years'
2 = ' 2-25-34 years'
3 = '  3-35-44 years'
4 = '   4-45-54 years'
5 = '   5-55-64 years'
6 = '     6-65+years';
run;
```

After re-running my newly spaced formats and adding the ORDER=Formatted option to the PROC LOGISTIC, I get the following class profile:

```
proc logistic data = cvd order=formatted;
class smoke age gender /param=ref;
model stroke = smoke age gender;
run;
```

Class Level Information							
Class	Value	Design Variables					
smoke	1-Current Smoker	1	0				
	2-Former Smoker	0	1				
	3-Never Smoker	0	0				
age	2-25-34 years	1	0	0	0	0	
	3-35-44 years	0	1	0	0	0	
	4-45-54 years	0	0	1	0	0	
	5-55-64 years	0	0	0	1	0	
	6-65+ years	0	0	0	0	1	
	1-20-24 years	0	0	0	0	0	
gender	1-Male	1					
	2-Female	0					

Figure 20.9

Now that the reference groups are all ordered as desired, I can proceed to look at the results.

Type 3 Analysis of Effects			
Effect	DF	Wald Chi-Square	Pr > ChiSq
smoke	2	15.2376	0.0005
age	5	115.6549	<.0001
gender	1	0.2406	0.6237

Figure 20.10

The Type 3 Analysis of Effects table tells me that both smoking status and age are significantly associated with the risk of stroke, but that after adjusting for smoking and age, gender is not. The odds ratio for a given relationship is the odds of y given x, adjusting for the effect of the other variables in the model.

Odds Ratio Estimates			
Effect	Point Estimate	95% Wald Confidence Limits	
smoke 1-Current Smoker vs 3-Never Smoker	2.097	1.443	3.046
smoke 2-Former Smoker vs 3-Never Smoker	1.399	0.998	1.960
age 2-25-34 years vs 1-20-24 years	2.874	0.353	23.398
age 3-35-44 years vs 1-20-24 years	4.919	0.634	38.174
age 4-45-54 years vs 1-20-24 years	8.040	1.071	60.350
age 5-55-64 years vs 1-20-24 years	22.762	3.138	165.125
age 6-65+ years vs 1-20-24 years	41.117	5.723	295.430
gender 1-Male vs 2-Female	1.076	0.803	1.441

Figure 20.11

Therefore, from this model, we estimate the following:

- Current smokers have 2.01 (95%CI: 1.44–3.05) times the odds of a stroke as compared to never smokers after adjusting for age and gender.
- Former smokers have 1.40 (95%CI: 1.00–1.96) times the odds of a stroke as compared to never smokers after adjusting for age and gender. (Note that because the confidence interval includes 1.0, this is not statistically significant).

- Those who are 65 years old or older have 41.12 times (95%CI: 5.72–295.43) the odds of a stroke as compared to those 20–24 years old after controlling for gender and smoking status.
- And so on.

Using the (REF=) Option in the Class Statement

Another option for controlling the reference groups in SAS is to use the (REF=) option in the CLASS statement. This option can be quick and easy but has to be repeated every time you model the data, which may be tedious when using a stepwise modeling strategy. The other downside is that IF formats have been applied to the data, the (ref=) option requires the format, not the underlying value.

For example, to set the reference coding for the preceding variables, I would need the following syntax:

```
Proc logistic data=datasetname;
class smoke (ref = '3-Never Smoker') gender (ref = '2-Female')  age
(ref = '1-20-24 years');
model stroke = smoke gender age;
run;
```

The formats listed in the quotations must be exactly the same (including capitalization) as those in your PROC FORMAT statement. Regardless of the approach, both should yield the same results. Order=formatted is perhaps less finicky, but its utility may be directly related to the number of variables you need to reorder.

Summary

Logistic regression is a useful tool for calculating the odds ratio and 95% confidence intervals in situations where the outcome variable is dichotomous. Logistic regression can calculate odds ratios for multiple exposure groups and odds ratios that are adjusted for a potential confounding variable.

Lesson #20—Study Questions

1. What is the value of logistic regression over simple odds ratio calculation?
2. What is the value of the CLASS statement in PROC LOGISTIC?
3. By default, SAS will model the probability of the outcome as what?
4. What are two methods for choosing the reference group for the outcome variables in PROC LOGISTIC?
5. Where in the SAS output can you find out which probability SAS is modeling?
6. By default, SAS will choose reference groups for the predictor variables based on what criteria?
7. What are two methods for choosing the reference group for the predictor variables in PROC LOGISTIC?
8. Where in the SAS output can you find out which reference group is being used for the predictor variables?
9. What is the format of the MODEL statement in PROC LOGISTIC?
10. How do you interpret the odds ratio from an unadjusted model?
11. How do you interpret the odds ratio from an adjusted model?

Appendix: Lesson Modifications for SAS STUDIO

Lesson #1: Getting Started with SAS (Studio)

Opening SAS Studio

To open SAS Studio, you must first

1. Open your virtualization software (Oracle Virtual Box or VMware),
2. Power on the virtual machine, and
3. Open SAS University Edition in your Browser.

These are the same steps you used when installing SAS University Edition. Return to the Quick Start Guide **(available here: http://support.sas.com/software/products/university-edition/docs/en/ SASUniversityEditionQuickStartVirtualBox.pdf)** steps 4 and 5 for more information.

Navigating through SAS Studio

When you first open SAS Studio, there are two basic sections.

Appendix Figure 1.1

In the center of the screen, you'll see your Program with three subtabs: Code (editor), Log, and Results (Output).

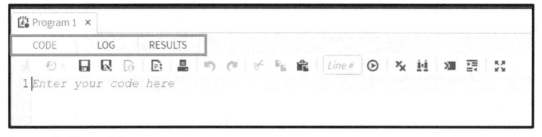

Appendix Figure 1.2

When you are in the CODE window, you can write your SAS Program as you would do in Base SAS. The CODE window is where you write all of your SAS code. This is where you tell SAS Studio what you want it to do.

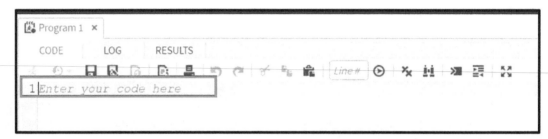

Appendix Figure 1.3

The Log

Appendix Figure 1.4   The log window displays information about how SAS executed any commands and about any errors. Blue or black text is (usually good). Red or green text indicates a problem with execution.

Initially, the results tab is blank, but it will populate with results when we begin running programs. The results tab works like the output tab in Base SAS.

On the left side of the SAS Studio window is your navigation pane.

Menu Bar:

Once you have written any code, the menu bar in the CODE tab looks like this:

Appendix Figure 1.5

Hovering over the buttons will tell you what each does. Feel free to click around and get accustomed to the different options.

From left to right, the buttons do the following:

- The running man (far left) is used to execute "run" programs or sections of programs
- Submission history—Displays program submission history
- Save—Saves file
- Save As—Saves a file in a specified location
- Program summary—Displays program summary
- Prints code
- Undo
- Redo
- Cut
- Copy
- Paste
- Line—Can navigate to a specific line of code by typing in the line number and pressing the arrow to the right
- The double XX = clears all code (Be careful with this one!)
- Binoculars = Find and replace code. This can be very useful for finding coding errors!
- The last three buttons are formatting options

Appendix Figure 1.6

In the LOG window, you have buttons (left to right) for Saving the file, Downloading, Printing and some formatting options.

Appendix Figure 1.7

In the results window, you have some very helpful options from (left to right) downloading the results as a HTML, PDF, or RTF file, as well as printing and some window formatting options. In Lesson 3, we'll discuss ways to save output in Base SAS. SAS Studio makes this much easier.

Appendix Figure 1.8   This navigation pane will be very important for accessing data in SAS Studio and for saving your programs, etc. This navigation pane replaces the explorer window in Base SAS.

At the top of your SAS Studio window, there are a few additional buttons that may be helpful.

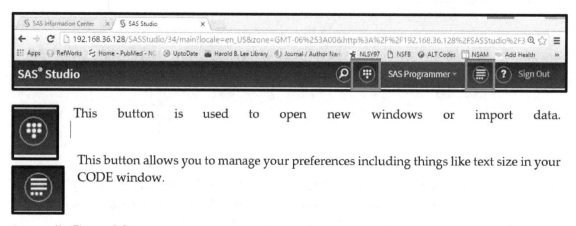

This button is used to open new windows or import data.

This button allows you to manage your preferences including things like text size in your CODE window.

Appendix Figure 1.9

SAS Libraries in SAS Studio

In most respects, SAS Studio and Base SAS operate very similarly. One key difference is the SAS Library. Since SAS Studio operates in a virtual environment, setting up libraries using the libname is less intuitive. I'll discuss a few ways to open your data using SAS Studio.

The Libname Statement Using Your SAS University Edition Folder

When you first installed SAS Studio, you created a shared folder in your virtualization software called "SASUniversityEdition." Within this folder, you created a folder called myfolders. This is your first data storage location in SAS Studio.

On my machine, you can see that my SAS University Edition folder is saved within my documents. You may have saved this somewhere else.

Appendix Figure 1.10

Double-clicking on the "SAS University Edition" folder and again on the subfolder "myfolders" will show me the files I have saved here. Anything I save in this file can be accessed by SAS Studio.

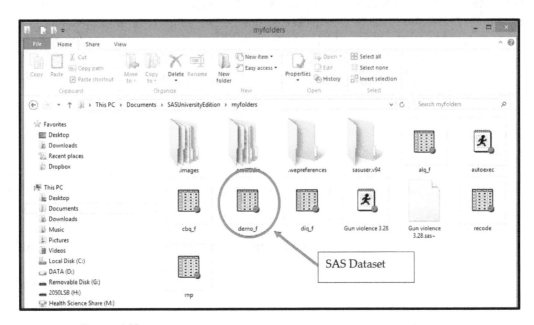

Appendix Figure 1.11

You can see that I have a few folders, some datasets, and two SAS programs (with the running man). While you are still getting comfortable with SAS Studio, I would recommend you save any data files you plan to use in this folder.

When you open SAS Studio, you should see that any datasets you have saved in your SASUniversityEdition/myfolders/ location appear under "My Folders" on the navigation pane.

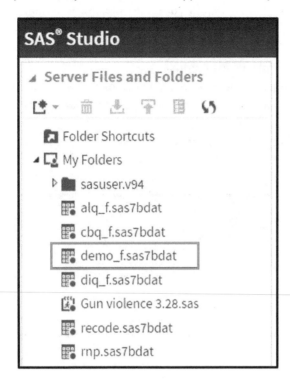

Appendix Figure 1.12 Notice that the data files that I saw earlier are also visible under "My Folders."

The libname statement is a SAS statement that creates the shortcut to the folder on your hard drive or removable media where your datasets are stored. This does not create a new folder; it just assigns an "abbreviation" to the filepath so that SAS can easily find your datasets. Then this "abbreviation," called the libref or the library name, can be used within the SAS editor to refer to the data.

The syntax for a libname statement looks like this:

```
libname libref 'filepath';
```

- Libname = the SAS keyword (light blue).
- Libref = the name of your library (letters and numbers, no special characters, and ≤8 characters).
- Filepath = where the folder containing the datasets is saved on your computer or flash drive.

Because SAS Studio uses virtualization software, the filepath to your data files is not as intuitive. For data files saves in your SASUniversityEdition/myfolders location, you will use the filepath **'/folders/myfolders.'**

To get to the folder that contains my datasets (SASUniversityEdition/myfolders), I would use the following syntax:

```
libname nhanes '/folders/myfolders';
```

Type the syntax. Highlight it and hit the running man button.

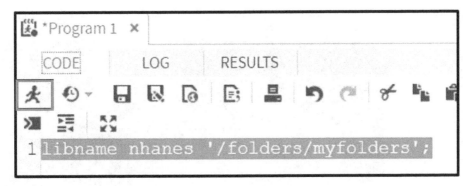

Appendix Figure 1.13

Note: I've used the libref nhanes, but you can use any libref you like as long as it meets the afore-mentioned criteria.

You can tell that your library has been assigned by looking for the following message in the log.

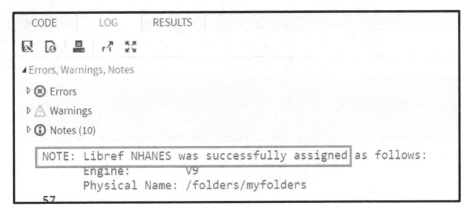

Appendix Figure 1.14

If you have an error, make sure that the subfolder within SAS University Edition is called myfolders. If it's called something else, you will need to alter your filepath. Directory paths and filenames are case sensitive.

Creating Libraries for Other Shared Folders

It is recommended that you use the preceding method until you feel very comfortable accessing data in SAS Studio. However, once you become more proficient in SAS, you may have more data files than you can keep organized in that single folder.

The first step is to add a shared file to my virtualization software. I will walk through the process using VMware Workstation. Refer to the quick start guide, step 3, for additional instructions for Oracle Virtual Box (http://support.sas.com/software/products/university-edition/docs/en/SASUniversityEdi-tionQuickStartVirtualBox.pdf).

Within my virtualization window, I will right-click on SAS and select Settings at the bottom of the menu.

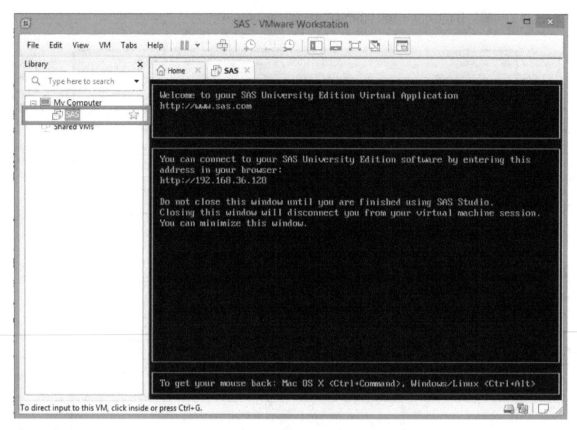

Appendix Figure 1.15

From the Settings menu, I'll select the Options Tab and the Shared Folders item.

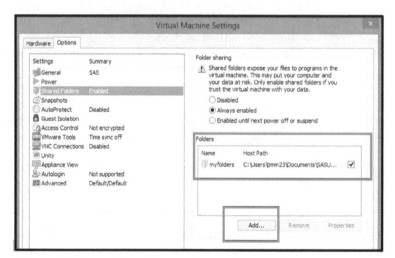

Appendix Figure 1.16   Notice that right now, I just have one shared folder. I'm going to add a second one in this example using the Add button.

By selecting "Add," the Shared Folder Wizard opens. I can then navigate through this to the folders I wish to access in SAS Studio.

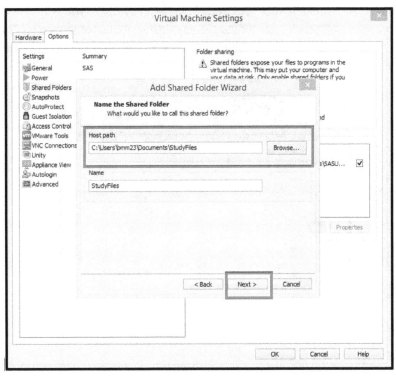

Appendix Figure 1.17 My folder is called StudyFiles and is saved in my documents. You can also access removable media through this means. However, you will not be able to use cloud files.

I'll click through the remaining prompts, being sure that the "Enable This Share" box is checked and that the Read-Only box is NOT checked. Once I have done this, I can see that I now have two folders shared with VMware. I can follow the same procedure to share additional folders if desired.

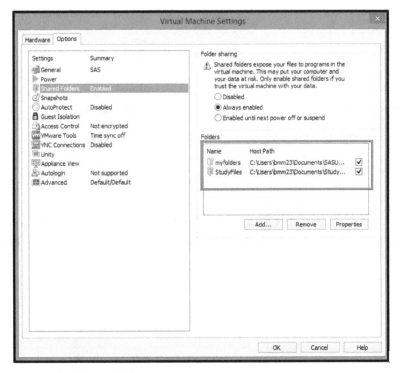

Appendix Figure 1.18

You can now open SAS Studio. If your SAS Studio was already open, you will need to refresh SAS Studio. When you reopen SAS Studio, you should see your other shared folders under Folder Shortcuts. If you do not see your shortcut, go back to your virtualization software and ensure that the folder is listed as a shared folder.

Appendix Figure 1.19

We can now use a libname statement as we did before.

```
libname libref 'filepath';
```

This time, I'll use the libref study (although you can choose what you like). Here the filepath is different than before. This filepath is '/folders/myshortcuts/file-name'. Directory paths and filenames are case sensitive. In my case, this means that my libname statement will read:

```
libname study '/folders/myshortcuts/StudyFiles';
```

Notice that because my Shortcut is StudyFiles (capital S, capital F), I use that capitalization in my filepath. Navigating down to Libraries, I can see that my library Study has been created and that it contains one dataset.

Appendix Figure 1.20

Creating Libraries with the Point-and-Click Method

You can also use the point-and-click method to create libraries. This same procedure works whether you are creating a library for your myfolders location or for another shared file.

Right-click on My Folders or the Folder Shortcut you want to set as a library (note: you must click on the folder, not the dataset). Select "Create" → "Library." The following window opens:

Appendix Figure 1.21

Enter the name you wish to use for your library in the box. I'll change my libref to NHANES, as I did before, and click OK. I can now view my files in the library navigation pane.

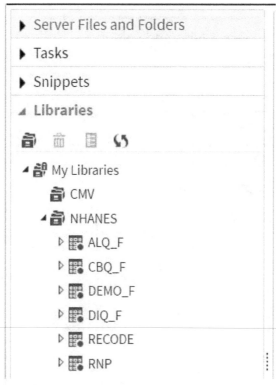

Appendix Figure 1.22

Saving SAS Programs in SAS Studio

SAS Studio will autosave your programs every 30 seconds by default. This is a handy backup. However, you should still make a habit of saving your program at the beginning of your session so that you can always locate it when you need it again. In the CODE tab, press the "Save As" button.

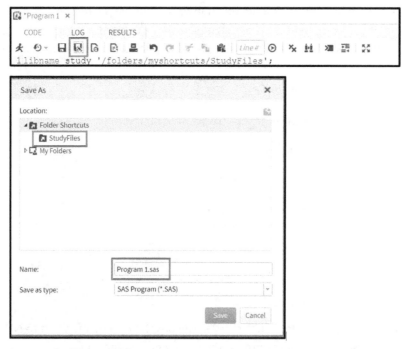

Appendix Figure 1.23

Highlight the shared folder that you want to save the program in (for me, it's either my StudyFiles folder or my My Folders location). Rename the program (by default, it saves as Program 1 (or some other number)) and click Save. Now I can see that the program is saved in my Study Files (a location known to me on my hard drive) and the name of the program is displayed on the top tab.

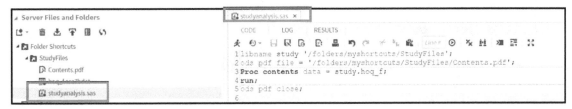

Appendix Figure 1.24

A Couple Other Notes about SAS Studio

In most cases, the remaining SAS statements will be the same in SAS Studio and Base SAS. Be aware of the following things:

1. SAS Studio on occasion will not color all of the text that is colored in Base SAS and will sometimes color things that shouldn't be. Don't be alarmed. If it works, it works.
2. While you are typing in SAS Studio, it will produce predictive text. You may find this useful as you initially begin using SAS. However, if you prefer, you can disable this in the application options.

Appendix Figure 1.25

Select Preferences from the Editor tab, uncheck "Enable Autocomplete," and Save.

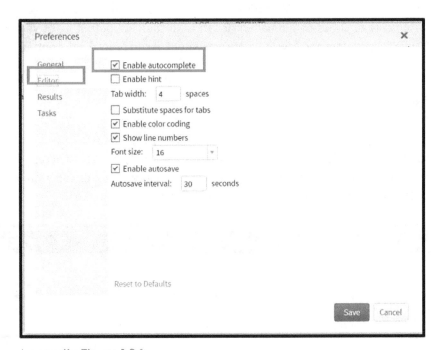

Appendix Figure 1.26

Where applicable, I will try to be specific about additional differences between Base SAS and SAS Studio.

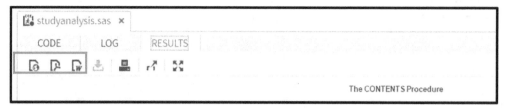

Appendix Figure 1.27

Lesson #3: Exploring SAS Data Using ...

Saving SAS Output in SAS Studio

The methods outlined previously for saving output files in SAS also works in SAS Studio. However, SAS Studio has simplified this process for you. From my results window, I can click on any of the first three buttons to save the data as an HTML, PDF, or RTF file.

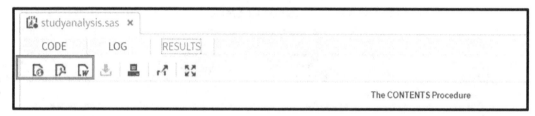

Appendix Figure 3.1

There is one important difference. SAS Studio will send this folder to your downloads. You can still use the ODS statements to choose a permanent location to save the output in, or you will need to open the SAS Studio saved folder and save it where you would like to preserve the output for later use.

Lesson #5: Entering Data into SAS ...

Using the Import Wizard to Convert SPSS, STATA, and Excel Files into SAS Studio

The Import Wizard in SAS Studio works a bit differently. You will only be able to access data files that are saved in your shared folders. Therefore, you will need to put any data files you wish to import into your My Folders location or share whatever folder they are in with SAS Studio.

Appendix Figure 5.1

I want to import the file Grades.sav, which is saved in my shared folder 440Data, into a SAS data file. To open the Import Wizard, I select "New Import Data" from the "New Options" menu.

Appendix Figure 5.2

To import, I only need to drag my Grades.sav file from the left navigation pane into the outlined area (or use the select file button).

SAS Studio will automatically generate the code to import the data.

Appendix Figure 5.3

By default, SAS Studio will save this data file in the work library with sequential files titled Import1, Import2, etc. You can modify this to save your data file in a permanent location.

I have a library called Class that saves to my 440Data shortcut. Under settings, I can click Change and select the library and rename the data file.

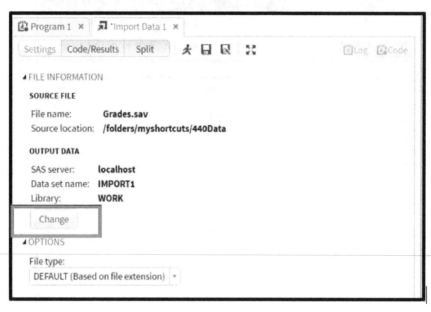

Appendix Figure 5.4

Appendix Figure 5.5

Clicking Save will modify the preceding import code.

```
 8
 9 FILENAME REFFILE "/folders/myshortcuts/440Data/Grades.sav"
10
11 PROC IMPORT DATAFILE=REFFILE
12     DBMS=SAV
13     OUT=CLASS.GradesImport;
14 RUN;
```

Appendix Figure 5.6

Now, all I have to do is run the code. I can see that my data file is saved under my Class library.

Appendix Figure 5.7

You can use the same process to import STATA or Excel files.

Lesson #6: Importing SAS Transport ...

Importing a SAS Transport File in SAS Studio
At this time, SAS Studio does not support SAS transport files.

Importing a Text File into SAS Studio
.txt files
When importing a .txt file into SAS Studio, you can use the same procedure as shown for Base SAS. Within your program, write your infile statements. The primary difference is the filepath. As we saw with libraries, because SAS Studio uses virtualization software, the filepath is different.

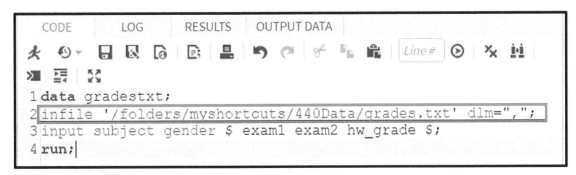

Appendix Figure 6.1

Note that as with all SAS Studio filepaths, I am using forward slashes "/" only. Also, note that I'm using the same /folders/myshortcuts/ initial portion of the filepath as I would for any Folder Shortcut in SAS Studio. The remaining filepath is as it appears in Windows with the exception of forward slashes instead of backslashes.

.dat files

The NSFG .dat file also imports the same way in SAS Studio as it does in Base SAS, again with alterations to the filepath. To import this data file, I will make the following changes to the input statements.

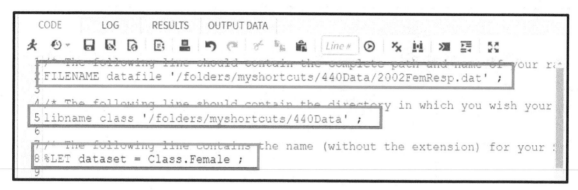

Appendix Figure 6.2

1. In the Filename statement, change the filepath to the location of the data. Note the forward slashes and that filepaths are CASE SENSITIVE. This must be a folder that you have shared with your virtualization software.

2. In the libname statement, change the filepath to the location where you want to save the data. Note the forward slashes and that filepaths are CASE SENSITIVE. This must be a folder that you have shared with your virtualization software.

3. In the %Let statement, I've added a libref (Class) prior to the data file name to ensure that it saves as a permanent file.

CPSIA information can be obtained
at www.ICGtesting.com
Printed in the USA
LVHW010012250720
661461LV00009B/68